The Book of One

Carlos B. Relano Jr.
The Book of One: How to save the world

Published by Spines
ISBN 979-8-90002-841-5

The Book of One

How to save the world

It is about alleviating poverty but I ended up saving the world after writing it.

Carlos B. Relano Jr.

Contents

Part Two
My Miscellaneous Poems

Preface

The past is just an illusion
The present is what we see in our souls
The future is what I came for in this world

For the West is now the East
And the East is now the West
The East has no choice
And so is the West
They have to integrate with the West
And compete, of course
In the End, It's a Win-Win Situation

As one, we are all perceived as unbounded
As a whole, we are all endless in projecting progress
As with everything, we are all one

We do not decide the future, we decide it for ourselves,
as we are all one, we are all like unbreakable rock

A Brief Summary

The book is about the world's salvation. The world needs help. It needs to heal. It needs a miracle. Though the world is getting better, it still has fractures that need to be resolved. For that, the world needs a guide. A guide that will provide light. A light that is a guide. A luminary. That will not be me. That will be society.

There are several reasons the book was made.

First, I found a way to alleviate poverty. Though it is simple, it is a complex process to fathom.

Second, I found other ways to fix the world, such as fighting storms, earthquakes, pandemics, and many more. They are primarily theoretical and may lead to my demise if it is found unworthy.

Third, I found ways to make developing countries investible. I wrote something about Central America, Myanmar, Africa, India, South America, and many more. I found ways to make the countries described worth living in, thereby leading to their development into developed nations.

Fourth, I found ideas and knowledge that are considered worthy. However, most of the provided ideas and knowledge are theoretical. I hope readers will find the theories remarkable, for they may lead to changes in the world. It will give society upgrades.

Fifth, society needs changes that can be addressed to improve them. I find it amusing that society is becoming more intertwined due to the technologies made.

Sixth, speaking of technologies, I provided technologies that may be deemed worth it. It may be considered a change in civilization. Remember, we are in a technological age. We are all bound by it. We can't live without it. We have to make progress out of it. Because of that, the technology chapters have a real meaning.

This is the brief summary of my book. Please enjoy reading it with pleasure.

Adfectus

I can Hear it... Smell it... Feel it... Sense it.... The feeling of absolution is now complete. Today, I can feel the hegemony of every country gleaming... for my tears have now dwindled... I have felt the rejuvenation... POVERTY IS NOW ERADICATED. THE WORLD IS NOW HEALED.

For our hope has been heard, and the mission for which we have been fighting is now accomplished. We are in absolute absolution, and it is now in contemplation. It is now in the midst... for it now verges... it now emerges....

IT IS NOW ONE

For we are one. We are many. We are a society that must accomplish greatness. This is the time to be resolute. It's time to achieve perfection. Time to accomplish interdependence.

For once, we are nothing. We are now everything.

For once, we live in darkness. We are now enlightened.

For once, we live in fear. We now feel freedom.

Together, we can accomplish anything. There will be none, and we will know nothing.

God chose me for his guidance and his almighty

THE BOOK OF ONE

THE BOOK OF MANY

THE BOOK THAT WILL ACCOMPLISH EVERYTHING

THE BOOK THAT WILL TOLERATE ALL AS ONE AND ONE AS ALL

FOR THE BOOK OF ONE WILL ERADICATE THE MEANING OF HARDSHIP

FOR THE BOOK OF ONE WILL ACCOMPLISH PEACE AND HARMONY

AND IN THE END, WE ARE ALL THE SAME

Introduction

The Book of One – the book that resembles unification, integration, and one direction. One world, one society, one vision, and one thought. It comprises several theories that encompass the actual movement of society through its economy, government, politics, and global affairs, as well as geography, futurism, psychology, sociology, and many more. The true meaning of the book is about who we have become and who we will become in the future. For 10,000 years, humans have evolved into one of the most intelligent species on the planet. We must comprehend what we were, what we are, and what we will be.

Can there ever be a unified world?

A utopian world is one in which human beings are equal in status and rights, and in which law and justice serve everyone equally. It is a world where poverty does not exist, yet the upper and middle classes live in harmony. It is a world filled with dreams and wonders, leading us to believe that everything can be achieved not through miracles but through the power of the mind and the progress of society.

Is Utopia a socialist message to the world?

I define Utopia as a message of equality, where the rich and the poor are almost the same. The poor are now middle-class, while the rich remain wealthy but humble and philanthropic. I know it sounds peculiarly intrusive, but we have to find a way to make inequality a solution to the problem we are facing today - the rich are too rich, and the poor are still poor. To fix that, read my book aloud, and you will find the real meaning of equality.

We are bound to this world. This is our aim. This is our right. We are dedicated to committing ourselves to this land, for it is the right way. Yet we are not doing it. Several anomalies, such as wars, famines, calamities, and destructions, contribute to such indifference. But the most daunting anomaly is war, and it will be seen as a challenging task that needs to be abolished.

So, what is my aim? What is my foundation?

The world is on the verge of another war today. It was not just an ordinary war but a world war. My goal is to end this war and promote peace, ensuring our survival and societal progress. My aim is to promote unity within ourselves to prevent humanity from being torn apart by the war that is to come. My aim is to live. My aim is to love.

The two World Wars of the 20th century led to hundreds of millions of deaths and the destruction of cities, and those who survived will never forget the daunting lives they experienced. Those wars cannot be repeated. They cannot be experienced again, for the next war will lead to our demise and the banishment of society. The next war will lead us to pure defeat and bitterness.

So, what is my purpose?

The world is becoming smaller and more accessible. In just one generation, a country with a vast population and abundant resources has become the second-largest economy in the world and the largest in Purchasing Power Parity (PPP)[1]. This is China. Its growth is even referred to as the 'great miracle.' For such growth is so daunting that other countries in the world will only look with jealousy, envy, and despair. Its development has made it the largest middle class in the world, comprising a staggering 400 to 600 million people. However, even with this middle class, it remains the second-largest economy in the world. The United States, being the largest, has the largest upper class in the world.

As history repeats itself, China and India are becoming more intertwined. The influence of the two is now widely regarded as one of the most remarkable comebacks in human history. However, something must be done so that the future is not disrupted. Something must be done to ensure uniformity. Something must be done to be one. Currently, the two countries conflict. In fact, the two countries will not be intertwined until both realize their importance for the future of humankind. With almost 3 billion people, the two countries must find ways to reconcile to ensure peace and stability not only in Asia but across the world. However, disruption will constantly interrupt its progress. That is why we cannot allow such things to cause dissolution.

The two nations had been among the most powerful economies for centuries until the West arrived and disrupted their progress. Too much destruction had led to the demise of the two countries. India became a colony of the British Empire, while China became a leading target of shame for several countries, which ultimately led to its decline and eventual destruction.

However, thanks to the technology available today, China and India are now at the forefront of becoming the leading nations of the 21st Century. Even though the West is disrupting China's economy with tariffs and sanctions, India is poised to get a sky-high GDP growth. The two countries are doing well and may do well in the future.

The Book of One is not just about these two countries. It is all about the intertwined world, the world that would lead into one. Yet it is not easy. It will be hard. Arduous work is necessary to ensure progress and maintain resilience. Yet all the countries must open. It is their only choice, especially North Korea, which has been so since the war in the fifties.

Let us also seek reconciliation with countries considered enemies of the West. Venezuela, which holds the world's largest oil reserves, must learn how to manage its valuable resources effectively. Iran, which also has vast oil reserves, must find a way to eliminate the sanctions. Cuba must open. Communism has failed, and they must realize it. These countries need help. I believe the BRICS are helping these countries. The West has to help them as well, for they are overdue in their insubordination. I think they want to reconcile. They want peace.

Europe's economy has been sluggish, if not stagnant, for years. That anomaly must be brought up as a priority. Otherwise, it will just be a continent that is free, yet no progress has been made. However, I must assess whether that is true, for I have seen promising progress in recent times that may be truly remarkable. It will be the chapter on Europe, of course. However, further research is needed to provide a definitive verdict. Europe needs to reunite. Great Britain needs to reconcile as well. The European Union must be as tough as an unbreakable rock, for it cannot deteriorate. It cannot diminish.

The United States also has an anomaly. Since the dawn of the 20th Century, the United States has been at the forefront of innovation and technology. It became the world's largest economy. Yet recently, something has been wrong. Since the Computer age of the eighties and the Internet age of the nineties, the United States has not taken any significant steps to improve its technological infrastructure. They were once unstoppable as they put their men on the moon's surface. However, nothing has happened for the last 20 years. There are computing devices, peripherals, the Internet of

Things, their gadgets, Artificial Intelligence, and applications, yet something is amiss. No modern technology has been introduced. There is robotics. There are space technologies. Yet, nothing was fully implemented. Nothing was fully introduced to society.

The most important aspect of this book is to introduce the process of human contribution to the world and its implications for the future. The most crucial contribution must be through human progress. Through human progress, we impose constraints and boundaries that must be broken. Unlimited boundaries must be conceived. Unlimited boundaries that will be unstoppable. Yet, with greed, envy, and the limitations of conservatism and progressivism, we found that progress was marred by fear and maliciousness. For some, the result leads to societal imbalance, affecting the world's progress.

But we must confront it for progress. It must be, for that is the only reason the United States can compete, especially with countries like China and many developing nations. It is the only choice. These are the only limitations we can perceive. For some parts of society, the true meaning of progress and innovation remains incomprehensible.

So, what have the developing countries achieved so far? What have been its implications? What has been its uncertainty?

So far, the progress of developing countries is unstoppable and intertwined. For the progress of the group is almost unbounded. Poverty alleviation is not just a success but also an unrelenting miracle. But more work needs to be done to achieve success. For example, Africa is truly unified with their pride and optimism. They are, but a medium-sized economic continent so far. However, they are aggressive. They will get to their goal. A Renaissance Age. That will be their aim. That will be their future.

While writing, I encompass too many ideas and pieces of knowledge to include in this book. I dwelt on many aspects that can help save the world, not only by alleviating poverty but also by addressing numerous other issues. Deep dive into this book. You will find these new ideas and knowledge profoundly encompass your mind with a deep understanding of the world and the universe itself.

The real meaning of this book is to accomplish what we call "WORLD'S GOLDEN AGE" or "GLOBAL RENAISSANCE." We are there. We are now more unrelenting. The only issue is that we are undermining the growth of developed countries, except for the United States, which is the only Western country experiencing growth. It is the only country that remains active. Japan, Europe, South Korea, and Australia are not growing as fast as they can. That must be changed. The United States must help achieve progress for Western countries. For developing countries like China and India, growth is so rapid that the rest of the world cannot keep up. The Western world must keep up with the times in the name of progress.

Once we achieve such fulfillment, we can not only make unrelenting progress but also reduce or even eliminate world poverty. Yes, it is almost impossible, but it is doable. This can be achieved through the careful distribution of goods and services, as well as the unrelenting progress of the global economy. So, can we achieve a Global Renaissance? Yes, we can. I am sure of it. I am dreaming of it.

And I will be perceived as the luminary that will guide the world through my light, for the darkness must be extinguished. Darkness must be upheld. Lightness must be perceived as victorious. We, the new society, are one. We are the new demigods, the new control of this world. The never-ending thoughts and ideas will be the actual progress. For we are one. Forget the past. We are now in the present. We are the future. We will be the redeemer. We are the actual builders of this world.

"And in the end, we are all the same. But we must live to be different."

We are perceived as just humans.
Yet our minds and hearts are bound for heaven
For that is what we are. That is what we will become

My life is already fulfilled, yet others have not started
We must all cross the finish line
For one day, we will become one

Part One
The World's Salvation

Chapter 1
My Struggle

My mind, my heart, my life, my soul—they are all in pain, becoming inadequate, and turning into dust. For what I believe is true has changed my innocence, my perception of the world, and the true meaning of love. For everything I see is false. It is just a broken mirror. All is nothing but malaise.

For what I see in this world has changed me forever—war, famine, poverty, hunger, hate, racism, sexism, and many more.

As a child, I have seen devastation that I cannot fathom. I have seen a child the same age as mine with earwax flowing out of his ears. He sells plastic bags. He doesn't even make a dollar a day. He can only eat once a day. He does not have a family. All he has is himself. And seeing himself is not just a misery. It is an actual truth. I hope it is just an illusion.

My life as a child in the Philippines was lavish. I live a life to the fullest. I can eat and play anytime I want. I go to a good school. I went to a place that only people with money can afford.

Yet, as a child, I already saw indifference to what I see as an illusion. For I see people with lives different from mine. I see poverty. I see hardship. I see people working hard, but at the end of the day, they can only get a small amount of income. As a child, I saw rich people getting richer while poor people getting poorer. But I am only a child. There is nothing I can do. There is nothing I can provide. All I can do is look at them and hope one day, they will become like my family. But I must realize the truth. They will not become like my family. That is because the system failed them, for they all live like animals competing in a very harsh environment. They are becoming rotten. They are nothing but rats—a rotten rat.

I am ashamed. I am angry. I am lost. But all I am is just clandestine because I am just a child. I am just a nobody. I can provide a small amount of money, but I cannot provide it for life. I cannot offer them lavishness. I can only offer them false guidance, tempting them to believe there is still a chance to live a good life. But there is no chance. It is just a false narrative.

Yet, I have seen one thing I do not see in my family. They are happy, confident, and full of intimacy. They still believe they can live life to the fullest.

That distraught me. It intensified my heart's desire to do something. My soul became radicalized. My mind became more intense. I needed to change my lavish life. Now, I have found an opportunity to change. I need to take advantage of that opportunity. I need to take a turnaround, for I have struggled with this mindset. I have struggled with all these occurrences. I have struggled with all these environments that surround me. I have struggled to realize that there is poverty. I have struggled.

But I must aim. I must find the right direction. I must follow my guts, for that is my reason. That is my fulfillment. That is my choice. That is my profound right.

So, tell me. What causes poverty? What causes individuals to live like it? What causes hardship? What are the causes?

There are several, but I can only provide the best and most

obvious answers—answers that can be taught in elementary schools. These include war or conflict, a lack of access to nutritious food, inequality, limited job opportunities, inadequate government capacity, and poor educational outcomes.

These issues have been addressed by governments, institutions, businesses, and not-for-profit organizations for a considerable time. It worked, but only in developed countries. Underdeveloped countries, developing countries, and emerging economies continue to struggle to find solutions to these issues. They are still struggling to find rejuvenation. They are still trying to determine what needs to be done. That is because corruption perceives them. They failed to make an effort to tackle these issues. Greed comes first before help. Malicious envy comes first before hope. Individualism comes first before society. This must be stopped. This must be tackled. This must be answered.

Yet, as I struggled to find clarity, I found myself ashamed, for these countries should have tackled poverty a long time ago. Yet nothing has changed. Nothing has been solved. It mesmerized me. It destroys my effort to tackle it.

But I must persevere, for this struggle will always be on my mind. It will keep me hanging. It will keep me unbound. It will be hard for me to cope, but I must solve these problems. I must find peace. I must TACKLE POVERTY. For this is my struggle. My aim is to change society and the world.

A broadened mind must be taught with a pure heart and soul
The effort must be strong
For its progress will lead to one

Life is short, yet an individual's ideas are infinite
It is a contribution that must be put into everyone's mind

Faith is a direction to the right path
Religion is a concept that led to a dead end
Divine Intervention is always the key to miracles

Chapter 2
My Worries

(ANOMALIES)

My worries, my guts, my diligence—they pertain to one flamboyant aspect—anomalies. These are signs—a warning, a prediction, and a problem that needs to be checked and countered. Every nation has its anomalies, and every country needs to be vigilant for warning signs.

Every country mentioned in this chapter has anomalies. Those anomalies might be easy to predict, solve, and counter. But they must be determined first, or it will be too late to tackle such an issue. Here are the anomalies that require attention and resolution.

United States Anomaly – 25 years ago, the United States introduced the Internet. It was a major success, yet for only a brief time, it was a failure, and only a few dot-com companies survived. For a few short years, those who survived thrived. Those who survived went on to expand into other regions and countries. The only problem is that the government and the military do not want to release another technological idea. For it may lead to another disaster, just like the Internet. That must be CHANGED. For 25 years, no technological ideas have been released due to fear or another disaster. Even though the internet has evolved into many ideas, such as the Internet of Things, Artificial Intelligence, Smartphones, Interactive appliances, and many others, it was primarily an innovation that served as an upgrade. Yet, we are long overdue for the next important thing. We need to release it now, or we will forever hold on to the technologies used as upgrades. For that, we must check the technology that might be useful for the country in the future. One technology is Space Technology.

Space is vast, infinite, and has seemingly endless resources. Those infinite resources must be mined. For example, asteroids near Earth have trillions of dollars' worth of minerals. Because they are near Earth's orbit, they can be quickly captured, brought near Earth, and mined. This will be a significant business opportunity, generating substantial profits.

The only issue is that we need to find innovative technologies to reach space as affordably and safely as possible. Rocket combustion technology is very deadly, dangerous, and costly. That needs to be changed. We need innovative technology that will enable us to travel to space cheaply, effectively, and safely. We need innovative technology that can help us utilize resources more efficiently. Those resources must be almost infinite because space itself is virtually infinite. I believe Earth's gravity and atmosphere are probably irregular. If that's the case, we might be able to go to space with ease.

Another technological idea that must be implemented is Robotics technology. It is a technology that may one day assist humans in their daily lives. They, of course, cannot replace humans as the next generation of workers. That cannot be done. That cannot be fathomed. Unemployment will go sky-high. That will lead to a disastrous American economy. It will lead to the destruction of society. So, Robotics must be implemented carefully. It needs rules, laws, and regulations. However, to truly implement robotics, investment must be cumbersome. Today, existing companies do not deliver the revenue and profits needed to make robotics a magnet for new and existing entrepreneurs. No robotic companies exist today that generate a billion dollars in revenue. That requires a radical change.

There are still technological ideas out there that may have been secretly implemented by the Government or the Military. We will have to wait if that technology needs to be carefully introduced to society.

Anomaly for Europe – the European Union has been dormant for decades. It has not grown economically. It has not grown exponentially. It seems the whole continent itself is on the verge of a dying flame. One issue for such occurrences is demographics. Europe is currently facing a higher-than-normal age due to a low birth rate and high mortality rate. Another issue is that the business environment is no longer competitive. Many European companies do not post higher revenue and profits these days. As a result, its stock portfolio is no longer attractive. Yes, the stock market has existed for hundreds of years, but it is no longer attractive. Government Debt is also high. As of the Third Quarter of 2022, the government debt-to-GDP ratio is 93.0%. However, unlike the United States, it has been decreasing gradually.

Anomaly for the Middle East – Iran and Saudi Arabia. The two foes will never stop hating each other. They are not particularly peculiar to each other, as their hate still occurs today. However, the issue might dwindle due to the recent peace talks between the two. However, this has been done before and has always led to catastrophe. The two countries have never held formal peace talks since 1979. And these recent peace talks may also fail. However, suppose Sunni and Shiite find each other a space for their conflict due to their culture, religion, military, economy, and society itself. In that case, they finally identify the differences and find a way to resolve them, paving the way for peace talks. Today, it is finally fulfilled. They are now intertwined and living peacefully. However, it is still fragile. The peace talks may break down. For they have been in a chaotic relationship for decades, and it may happen again. However, I have a dream that the peace talks will finally last. We just need to be patient.

India and Pakistan – two nuclear nations. Ready to implement their ICBMs because they believe that they will fix the conflict they have in Kashmir. Although the conflict has subsided, we must recognize that the issue remains unresolved. Radicalization in Pakistan has been subsiding, yet there are still extremists who need to be contained. The good thing is that the countries are doing exceptionally well with their economies, and employment is being provided for the people on both sides of the border. I hope that one day this will lead to peace and prosperity for both sides.

The betrayal of one might be the betrayal of many
A betrayal of many will lead to catastrophe

My respect for you may be unlimited
But provide me with unlimited insult
And it will limit to dire consequences

Chapter 3
My Mirage

(The Complex Thoughts)

Middle Class – the key to Economic growth. The key to improving society is that it is the key to eradicating world poverty. They are the key to the development of every business, organization, government, and many more. They have the power to make a country better or destroy it. It is an intimidating class because they indeed demand more consumption, and their usual situation is that they have enough money and wealth. Governments around the world can provide that money. They can provide that wealth. They must give it through a so-called basic universal income. An income is offered to everyone to ensure that their basic needs and desires are met. The middle class will not be created with that. But it is a start. We do need more. We need a compromise to create more middle class. As I said, they will drive the economic growth not only of the country but of the entire world itself. So, how to start a middle class?

We must concentrate on a country that provides the largest middle class. And that is China.

As I mentioned in the previous chapter, China's middle class comprises approximately 400 to 600 million people. They are the primary key to driving the world's growth. And the world was never the same, thanks to China's middle class with their hard work and dedication. In the 1980s, China had the largest poverty class in the world. Thanks to economic reform and development, the poverty class has significantly dwindled. And indeed, it created the biggest middle class that the world has ever experienced.

So how did China do it? Is there a model for the world's growth?

It is not easy. It is not radical either. It is up to you to determine if it serves as a model for the world's growth. For sacrifices have been made for such growth. Yet, with that sacrifice comes fast, destabilizing, yet productive growth. One sacrifice Chinese people must make is to relocate. It is indeed bad, but most, if not all, usually move to a better place with incentives, such as government payments for buying their land and better benefits. The original place they lived in was being built with high-rises, office buildings, and housing.

Another sacrifice the Chinese mostly make is that they tend to save a lot and spend little. That is indeed the reason for its economic growth. However, it is not the right way. For spending indeed is more important than saving. Recently, in the First-Quarter GDP report for 2023, China recorded a 5.8% growth in consumer spending[1]. That is a good change indeed. It is another good start for the country.

So why is China?

China proves to the world that anything is possible. Its GDP in the 1980s was worth $202 billion. For the year 2023, according to China's National Bureau of Statistics, the country's GDP is worth $18 trillion[2]. That is the most significant growth ever recorded in human history. It took the United States more than 60 years to get into that path alone. Such an accomplishment should be partially emulated by other countries, particularly developing ones that need a growth model for their expanding economies. However, the Purchasing Power Parity (PPP) of China is larger, at 34.54 trillion US Dollars. That will make it the biggest economy in the world, surpassing the United States' GDP PPP of 24.62 trillion US dollars, also in 2023.[3]

So, can we use China's growth as a model for the rest of the world?

It is hard to tell, but it is up to you to know if it is appropriate. That is because China uses radical methods to ensure its growth. Much of the world is progressive or liberal in orientation. They concentrate on their growth using conventional methods. Methods that are liberal models. For example, China builds cities and urban areas. The rest of the world can only partially build cities. China can move hundreds of millions of people from rural to urban areas. That is quite phenomenal, yet as a progressive, I would find that disturbing. For moving people requires a lot of challenging work. They need food and shelter. They need privacy, healthcare, and schools for their children. But can it be done? They already have. The urban population in China accounted for 66.16% of the total population in 2023. In 1980, it comprised only 19.39%[4]. The metropolitan population is expected to grow further in the coming years.

So why live in urban areas?

It is the most feasible way to ensure individual capabilities can be used in one place. Collaboration will be imminent for ideas that are indeed needed to cultivate urban growth.

Can a progressive country make such moves?

Yes. We have no choice. Urban growth is the main factor in economic growth. It is the only choice for such concentration in one place, making it easy for businesses and organizations to thrive and succeed. We need to persevere because we must concentrate on our growth in one area.

So, what are the employment numbers needed to ensure a Global Renaissance?

Just a slight majority. Approximately 3 to 4 billion people out of a total population of 7 billion. That is quite huge indeed. Yet, it is doable, feasible, and tolerable. Such numbers are indeed needed, as the rest of the employment will grow with that number. China has ensured the emergence of a 600 million middle class. The rest of the world must concentrate on providing actual numbers. As I mentioned, the middle class is the key to actual employment. It is the key to economic growth. For China's growth provides a labor force of almost eight hundred million people. It needed more. Four hundred million more. India needed one, too. It has a labor force of more than 400 million. It requires an additional $800 million to ensure economic prosperity. The United States has 165 million. It needs eighty million more. This is quite similar to other countries, especially major economies.

So, why am I asking for a larger labor force?

That is because basic income can provide a middle class and a labor force so strong that it will create such an occurrence.

But where to get that money?

We have no choice. We need to encourage wealthy individuals and large corporations to pay higher taxes on the wealth they have accumulated. It is calamitous that the wealthiest 1% receive twice as much as the rest of the world combined. While the rest of the world gets poorer, the rich enjoy a lavish life. I know it is not adequate, but it is indeed immoral. While people live in the streets begging for food, the rich beg for more yachts and mansions. Such a disturbance should not occur, for we are human, and inequality is inherent. Inequality has been a disturbing phenomenon since the Industrial Age of the 19th Century. For that is the age where people finally accumulate wealth so meticulously that it is hard for the government to redistribute wealth to the impoverished. Yes, the rich during that age accumulated so much wealth that it was so vast and destructive that the poor and unemployed could watch in horror and envy. However, the economic system at the time was amateurish and inadequate. Unprecise and undesirable. With that, the government learned. They finally found ways to redistribute wealth and fight inequality. Of course, the wealthy will always accumulate more money, but the impoverished can finally become part of the economic system. For they need to collect and redistribute wealth. For they are required to provide more jobs. For they are necessary to help the country grow and improve the economy.

Yet, a lot of fighting is still needed.

However, not all countries can provide a basic income for their citizens, which is not a fundamentally sound principle. Most countries today are developing and emerging. They cannot offer such a basic income. So, what must be done? Simple. It has not been created yet. The most essential thing for such a country is to provide employment opportunities. To do that, the government must create incentives for businesses. The government must provide training for future workers, schools for students, colleges and universities for college students. However, we need to do it fast while businesses are thriving. For once an economic bubble is created, it will be hard to stop it. Yes, the financial bubble is ensured in new economies. Once it bursts, it will be disastrous. Yet countries today are resilient to that. They know how to cope with and solve such occurrences. For they will know it will happen. They must find a way to find a soft landing, for a hard landing is the one they must avoid.

Another way for countries to become wealthier is through investments. They can be direct or indirect, but they must be welcomed regardless. And they must be used well. Of course, businesses depend on them too much, but discipline is the key for such a business to grow. For a developing country or an emerging economy, that occurrence must be fulfilled. Investors are humans, too; they must trust the companies and ensure honest work from them.

The last and most crucial factor for driving economic growth is eliminating corruption. Corruption is the most disastrous and sickening of all government factors. It is the biggest reason some developing countries stay developing. That is because corruption is so rampant that no one can stop it. In some countries, it is even the norm, and no one in society can stop it—only the people. However, people can also be manipulated. There is another answer to that.

Lee Kuan Yew provided such an answer that it works well for his country, Singapore. Thanks to Lee Kuan Yew's leadership, reforms, and many other initiatives, the government is now one of the best and most livable in the world. One aspect that he must battle is corruption. Yet he found a way to fix it. How? By providing higher income to the politicians. Singapore's politicians have one of the highest incomes in the world. And corruption is so rare that they forget such words in their mind. Corruption rarely occurs in Singapore. It never occurs. That's because once found, the officials responsible for it will face enormous punishment. The punishment is so severe that the world will look in fear. But that is how it works in Singapore—the most disciplined country in the world. The most feared for its justice is harsh. And it never chooses which one to punish. Everyone will be treated the same.

Lee Kuan Yew passed away in 2015, but his legacy has provided the most progressive yet intoxicating system in the world. Singapore is now fundamentally sound; however, you must follow the law, and you will be dealt with accordingly, with a handshake and a head bow. Thanks to his leadership, Singapore is now one of the most important cities in Asia and probably the world. And someday, we may need a Lee Kuan Yew moment. We will need his ideology.

To be here is to show your presence
To be nowhere is to show your irresponsibility

The devil will always be in our presence
But He will always be in our hearts and souls

To be an Ant is to prove your individualism
But nothing can stop a group of Ants, for they are colossal

Chapter 4
My Philosophy

(My Beliefs, Emotions, and Undaunting Theory)

I was once in dormancy. I have been living in a depressive state for my whole life. Living in a discretion, my life has been in dementia. A dementia of knowing nothingness, knowing the inappropriateness, and knowing what reality is. I am inadequate in my perception of life. My childhood had no purpose. Had no sense of direction. Living a lavish life will not provide any fulfillment. Living in undefined surroundings will not suffice to make me happy, for it is all the same. For I live in isolation, desolation, and an unfair environment. I live with no one. I live in a quiet, destructible, and immersive home.

Although I have a sister, we don't get along. We play, but we do generate childhood malignancies. We devour happiness, but the relationship is not fruitful. We are a bunch of imbeciles in an unliving manifestation. But we have to adhere. We have to mourn our intimacy. For we are children, but we are not. For isolationism develops unsatiable behavior. It creates an undignified life.

Yet we were bestowed upon our faith. We were Catholics, and we admired the tradition. We were timid, but society was engaging. We live in admiration and with no discretion. Fear was not in our minds. For innocence made us formidable. For it was an unusual childhood. It was an extraordinary life.

But more importantly, my childhood bestowed upon me with knowledge. For I was surrounded by books that defined me. For I love them. I live with them. And I developed a prognosis of information. Yet I remain scanty, dormant, and in self-isolation. I live in depression, and I live an unmanageable life. This "was" my whole life.

This is me now.

A new world came to me—new surroundings, a new culture with diversity, and a society with confidence—a society full of imagination. But that was back then. The environment changed, and the society changed. However, I had to adapt to the change. I had to alter my determination. I had to adhere to my beliefs. I had to develop a Philosophy.

Even as a child, I knew reality was a devastation that must be contained. For I know my choices are flourishing, but others' choices are not determined by it. It is determined to be the only choice and no other. And they have to live with that choice. And sometimes they have to live with no choices at all. But they still believe in enlightenment. For they believe... and they will live.

"For I believe"

I live with the determination that all societies will one day live an unexceptional or lavish life. No one is left behind, for the impoverished are impoverished no more. All of them will live a life of importance, moderation, cooperation, generosity, determination, advocacy, benevolence, diversification, equal reputation, confidence, and, most of all, humility.

"For I believe"

I wish that one day all countries would unite for one cause – in the name of progress. I believe that with technology and innovation, humanity will flourish and enhance its lives. They will give utmost importance to their reputation, for they are determined that one day they will provide unimaginable ideas and colossal creativity, ultimately establishing their own legacy. For they believe in it. They believe such things will exist. An intertwined world is a future I am determined to fight for because it will help combat poverty and shape the future for the next generations. With countries of integration, such occurrences will one day become a reality, and wars and conflicts will be a thing of the past.

"For I believe"

I believe in technological miracles. I think we can do an innovative miracle that is proven to be out of this world. I know that through cooperation and competition, we can devise a way to create a very advanced product. I know Technology companies are the salvation of a macroeconomy. That is because I believe they hold the utmost importance in macroeconomic structure. They have the key to humanity's future, and I think they can provide enormous jobs. Technology companies hold the future and are willing to give that future to make a change.

"For I believe"

Gifted and geniuses cannot just be taught. Some are born that way. Some do not even know they have it. We must exploit that. We know many of them are living in dormancy or are probably living in poverty. We cannot waste such skills and talents. That is a disaster that we cannot comprehend. It must not be allowed. We humans must find a way to help such magnificent ones. Of course, education is one of them, but some are shy or afraid. We must provide a forum in school to exploit them and galvanize them. Encourage them to speak, and one day, they will contribute significantly to the discussion and, eventually, to society. Give them a chance. Even if they are in the future, they are willing to help and be recognized. Such talents are wasted if we do not allow them to be tolerated.

"For I believe"

I believe that one day, poverty will diminish. Such an obscure idea will not occur today, but may occur several decades from now. Many countries, if not all, are fighting poverty right now. With the power of technology, we know how to work, compensate for checking, provide solutions, and take action to help the impoverished. We can do it. The Government is the primary contributor to such a cause. However, the upper class can be a significant contributor as well. We have to find a way to help them unite and work together to help people experiencing poverty. I believe Greed is a trait that can be eased. Of course, it can't be diminished, but we can find a way to relieve it and encourage the rich to work together to minimize greed and embrace generosity. Such traits can be helpful for the 1% who control almost 50% of the world's wealth, and we need to find a way to ensure they contribute to society. The middle class has to find a way to contribute to wealth, not through their wealth itself, but through their actions. The middle class can provide food, shelter, and everything else to the impoverished, and with that help, a cause is established.

"For I Believe"

I believe the mentality of an individual and society can change. I feel such traits can be diminished. For example, China and its people used to think of extreme socialism or communism. Now, they still believe in it, but they are willing to work on it with capitalism. It has proven successful and has become a win-win situation for the country. Iran, Syria, North Korea, Lebanon, Palestine, and many others need to change their mentality as well. I know it is difficult to comprehend such a mentality because their enemies are trying to find ways to take advantage of the situation. And with the help of the rest of the world, their enemies are winning. If these countries find a way to open themselves to the world, and the world is willing to stop the sanctions, a compromise will be reached. They do not have to demonstrate their military; instead, they need to take a reverse course by providing an open society that welcomes local and foreign investments (FDI) and becomes more innovative and business-oriented, especially in manufacturing. They have to assert openness to government transparency as well. They must be willing to find ways for the world to trust them, and transparency is one such way. Such discipline will work and be a win-win situation for both sides. For their enemies will slowly trust them and start to earn their reputation, and probably, one day, they will not be enemies anymore. Such philosophy is probably obscure today, but one day, it may work. We have to believe.

"For I Believe"

I believe that the past is just an illusion. We always make mistakes, and that will always have to be confronted with; find a determined solution, try to vanish it in your mind, take responsibility for such actions, be willing to go on with it, and always find ways to make choices, actions, responsibilities, and capabilities for your future. The future is always of the utmost importance because it determines who you want to be, and one day, you will be that person. You have to open yourself to the light and focus on it. If you can, pray to God, for he will provide guidance, help, and determination. Despite how good and talented you are, you remain vulnerable if you lack focus. But God will give you the proper emphasis if you recognize him and are willing to open up to him. Just keep the faith, and he will provide you with enlightenment.

"For I Believe"

Discipline is a must, and it must be allowed to flourish. Such discipline is needed to ensure society's survival and prosperity. Discipline is the key to why some countries have become some of the best places to live. Sacrifices are another trait that must be taught for society to function effectively. Both traits must be intertwined. This is one of the reasons countries like Singapore, Taiwan, South Korea, and

Japan have become some of the most developed countries in the world. However, many of them started in a dictatorship. I do not believe that a dictatorship is allowed to ensure prosperity. That is because we have technologies today that can enable us to provide discipline to society without using an iron hand. We can assert "limits" on society without controlling it with these technologies. Such limits are necessary because they will not lead to rigorous inaction and irresponsibility that could harm society. Such technologies are of utmost importance and will help society flourish. We have to find ways to use them. However, such technologies cannot be abused. Just don't.

"For I Believe"

I am deeply knowledgeable. Yet, I used my knowledge as a chess piece. For the game of chess has pieces with unique movements. That's how I use my knowledge. I interpret the knowledge I gain and use it as a puzzle to determine if it will work, provide more knowledge, and ensure it will not break apart. Such puzzles will be enhanced with more knowledge until I find the determined solution or stature. I used different branches of knowledge such as Economics, Geopolitics, Sociology, History, Geography, Technologies, Visions, Futurism, and many others. By today's standards, Economics is the key to a determined future. In my philosophy, it needs to be enhanced or combined with other branches to identify the right puzzle and determine the correct answer. But everything has its limits. And I call it Limit, which is usually associated with a conservative-minded person. Such actions need to be determined and studied with precision to prevent future mistakes. Like the mistakes we made on the Internet.

"FOR I BELIEVE"

Our educational institutions have excelled in teaching, knowledge, professionalism, and engagement with the world. The professionals with these traits become more isolated from other sectors of education. For example, an engineer is so technical in his words that a more professional person, such as a doctor, will not be able to understand his meaning. Thankfully, a doctor is always engaged with patients who need more information about their health, so he must maintain a simple understanding. But his words and voice are always technical or medically technical to his fellow doctors or nurses. So are lawyers, physicists, chemists, and many other fields and probably, or unfortunately, STEMs as well.

When I was in my previous Information Technology job, I was engaging in technical language and terminology. However, something in me wants to make it simple. I want people to make things simple. People need to know, engage, and probably provide something otherwise. People with knowledge do not need to share it extensively. They have to be simple. They have to talk like an average person. That's because others want them to know as well, not as a professional but as an engager. I am not trying to change the educational mentality. I want them to engage with the rest of the world.

"FOR I BELIEVE"

For the remaining chapters, emotional writing can't be provided, for such content requires technical intelligence and skills to ensure understanding. However, some viewers cannot comprehend such technical skills due to the complex and rare words that they use. Such a subject must be enumerated not in a technical aspect but in a perceived subject of simplicity. That's the only way readers can understand such subjects; many can, but some won't. For people who possess intelligence that is far more complex, they must ensure a complete understanding of the subject. To summarize it all, and I will provide it again, simplicity will always overthrow technicality. Simplicity will be my approach to writing about subjects that are too complex or complicated to comprehend.

Like smoke getting out of the chimney, it disappears quickly into thin air
But it always leaves a mark on the chimney
And that mark will increase slowly but surely
Just like reputation
It leaves a mark, but it always makes sure

Chapter 5
The Ultimate Order

(Complex Ideology)

Why Ideology? Why is there a need for a guide? For a luminary? These are the thought processes that need to be ingrained in our minds to ensure we accomplish what needs to be done, from the beginning to the finish line. We must provide ourselves with discipline, cooperation, participation, and resiliency to ensure that the ideology that we need to provide is enhanced and intimate.

So, what is my ideology? What is my inner peace? To provide such an ideology, you need to know the thought process that will alleviate you, help you enhance your mind, ensure continuity, and introduce thoughts with quality and guidance. To seek such an ideology, you must know yourself, your environment, society, and probably your love partner. For they will help to guide you, enhance you, and make you a better individual. He or she will provide you with love and intimacy. But this is not all about that. It is all about ideology. We, as people, must know what ideology we are after. What we are all about.

My ideology is liberalism. My ideology is democratic. My ideology is to alleviate poverty, help the poor, address inequality, and be a luminary. For I believe I am the guided light. For I wanted to be guided myself. I am liberal, and that is what I want to be. I am liberal, and that will not change me. I am liberal, and that makes me a person with a guided right path.

I will help people experiencing poverty, alleviate poverty, and guide them. I will help them. For that is my ideology. That is what I want to be. Do not provoke me, for I am determined. Do not shake me, for I want to be guided. Do not prove to me because I am proof enough to provide this ideology. Do not even change me because I cannot be shaken; I cannot be touched, for God himself has guided me. God taught me about this ideology, for God is my true ideology. God is my true guide.

So, who am I? Why am I talking about ideology? Why am I being a philosopher in such a manner? Why am I writing this book? I know it is all about alleviating poverty, but there is more to it. Something needs to be accomplished, and something needs to be open to society, something that is beyond my control.

I believe there are poor people out there who are gifted, ingenious, and have an IQ that is so unimaginable. I think those people are here. Some live a good life. Some live a lavish life. Some are living in a dream. But there are just those who are unfavorable. Living an impoverished life is disastrous. Living an impoverished life proves to be worthless. Because of that, they cannot prove their gift. They cannot prove that they are ingenious. They cannot prove that they can change the world, for they live in depression, envy, and anxiety. They live in an unfavorable inequality. I am writing this book to help them. To open their mind. To bring them into society. In the end, society will love them because they will bring fruit to society. They will make society better.

So, what am I trying to say?

Like what I am saying, depression, envy, and anxiety will close your mind. It will not open your inner thoughts. You will just be dormant. You will be a disaster, for you are nothing but a nobody. You are just an average, everyday person waiting to die, for you to be proven to be nothing. I must change that. I must ensure that there are people out there who can change society. To prove that I must write this book is to prove what is worth. To prove that nothing can change my ideology. That nothing can change me at all. For God is with me, and he will be with me eternally.

So why liberalism? Why not others?

Liberalism has been proven to be the right ideology to alleviate poverty. They are not as extreme as communism and socialism. They are not extreme at all, as they help people experiencing poverty with the proper guidance and manners. What I mean is they do it the right way. They ensure that people experiencing poverty do not depend on the government. They ensure that people experiencing poverty are not considered lazy. They ensure that people experiencing poverty are part of society. They ensure that people experiencing poverty are provided with good jobs and money. They ensure that people experiencing poverty will live a better life and eventually become middle-class.

In return, I want you all to know that I am not being sarcastic. I want you to know that other ideologies will work as well. My ideology was conservative when I was in the Philippines. But it all changed when I got to the United States. During that time, the President was Bill Clinton. And I was amazed by him. He was such a good leader. It would be an actual model if not for this scandal. But that did not change me because he was still a good President when I was there. He proves this with his qualities, personality, leadership, guidance, and, of course, his ideology. Though the scandal made him mediocre, he was resilient, motivated, and self-disciplined. For that, he won against all odds. He won against the many people who hated him. He won, and that proved to be a Good Leader. With him, the budget surplus was never a dream again. It is a reality that must be adhered to. He made America the true envy of the world, the unipolar power, and the wealth so vast and too abundant that it may take another generation or two to be in that environment. I experienced that period. I want to experience it again. I am hungry and thirsty for it. I am just too motivated. So, I want God to help me, guide me, and provide me with a dream and a reality. Please provide me with an ambiguous thought process. To prove that I am a luminary.

I was once a disaster, but you gave me a heart and soul
You have proved to me to be something. I will prove to you what I have become
I will prove to you my true self

God gave me grace, and God gave me enlightenment
God will always become my wonder
For I am dignified, I am becoming a part of his heaven

Chapter 6
Flamboyant Society

(The Group of Grandeur)

The society, once just a group of villages working together to survive, is now a grand complex of diverse people with different religions, occupations, consumption systems, military power, economic influence, technological power, and more. The society of today is just vast and complex. Industrious and resilient. They are the fundamental core of the community. For once, the culture was once indigenous; it is now dynamic. War and devastation have always put down the core of society. But society is resilient. It is always trying to recover. Always trying to improve. It is always trying to innovate.

Society today still needs significant improvement, especially in collaboration, networking, and accountability. Society today is intertwined with the world through trade, business integration, industries, and economic power.

So, what am I talking about? What is my core? What is my

dialogue? Why am I talking about society? Why am I talking about the community?

I want a society that works together, helps each other, and collaborates to find ideas and creativity. I want a society that knows how to integrate, ensure innovation, and make miracle inventions. Society today is not influenced by it. Sex, drugs, and violence influence it. It is influenced by people with money and power. It is influenced by religion and the faith-based community. I do not mean to make religion a bad influence, but I want to limit its control over the state and faith must be separated. That is how the United States, which is my country, was founded. That is what I want every country to be based on.

But my main goal is to make society work together to tackle poverty. I want society to ensure that every individual has a job or a career. I want to ensure that everyone will have food and shelter. I want everyone to live a good life. A society that is not only based on religion but also on science and technology. I also want society to integrate to seek what I call a utopian world. A utopia that has no definition of poverty, impoverishment, poor, or inadequate. What I want is a society of the upper and middle classes, with no lower class. I want a society with good inequality. I want a society that understands the meaning of sharing and giving, rather than greed, jealousy, and envy. It will be hard to start this kind of community, but we must work together to provide that kind of society. Because I want grandeur. I want harmony. I want cooperation. I want peace and order. That is what I want a society to be all about. That is what I want a society to become in the future. For the future is promising. The future that God can bring. A future that people will accept. A future that society can manifest. A future that has no poverty.

I was once a nomad. A wanderer. For my society back then was rude and cruel, bullying and reckless. I was devastated by that society. I was overwhelmed by their attitude. I was devastated because they wouldn't help me; instead, they just let me rot. Even though I was living a lavish life I was mocked and insulted by society because of their jealousy and envy. But it is fine. I understand them. However, they should have given me a chance to integrate with them. But they did not. They have money, power, and influence. And there is nothing I can do. The only thing I can do is to survive.

So, I moved to a new society. The United States' society. A more advanced and flamboyant society is emerging. A society that is hard-working and has an abundance of money. A society of innovation and a business-savvy environment. This is a society that I would love to be part of. A society that is blessed by God. That was in the year 1998. At that time, the Internet, also known as the World Wide Web, was still in its infancy. However, it was a society that needed reform because the country had just become unipolar. The country just won the Cold War against the might of the Soviet Union. And the American people were delighted with pride, joy, intimacy, love, and patriotism. It was a great society. It was the model and envy of the world. For God made it as the society. God integrated it as one united community.

I hope that God will give me the power and influence to teach people that an integrated society is necessary. I hope God will let me be a luminary to society. I want to guide them, change them, influence them, and let them know that poverty must be eradicated, and hope they can change the world for the better. I must tell them that that is my ideology. That is my aim. So let me tell you one thing. You are also a part of society. Be like one. Act like one. Aim like one. For you are a human being. You are also a luminary. God brought you to this world to be part of society. God gave you the knowledge to help society. You must help them. Help them at will. Help them become one.

I painted you to be my Mona Lisa
I painted you to be my core
Paint me too for I am the center of your painting
I am the center of your world

God is my shepherd
For I am his sheep
He will guide me to the right Pasteur
He will guide me to Heaven

Chapter 7
World's Golden Age

(Grand Renaissance)

Grand Renaissance. World's Golden Age. That is my aim. A reality that I want. A reality that is a must. The Renaissance was once a reality perceived by the Florentines and acknowledged by the rest of Europe. It was a period of never-ending accumulation of wealth, and it became the phenomenon that made Europe the true Envy of the Western world. Interestingly, before the Renaissance Age, a significant disruption occurred in Europe, leading to a disaster so enormous that it became the worst pandemic in the world. This is the Black Plague. The plague that changes Europe and the world forever. It changed European society and the world together. It changed the thoughts of the Europeans. It brought them disaster and dire consequences that changed their thoughts. But they turned, reversed course, and flourished. I do not mean to say that a plague can change society, but it changes the thoughts and perceptions of people. For they suffered, and the result was incomprehensible. In short, it made them better. The result was the Renaissance Age.

So how can we begin the World's Golden Age? What can start a Global Renaissance?

It is always the people. People of different diversities, cultures, communities, societies, ideologies, and class statuses. It is always their perception of the world. It is the perception of reality. And the fact is, people need knowledge, hard work, and resiliency. The World's Golden Age needs this kind of people. They need a dream of this kind that will become a reality. But to galvanize the people, one must guide them to ensure the integrity of the period. And that is my aim. That is what I want to perceive, for I want to guide the people to ensure that no one will be left behind. No one will be disrupted. Everyone will accumulate wealth. Everyone will be either upper or middle class. For the lower class will vanish. They must and they will be. For all will flourish. For all will become one.

Of course, to start one, the economy must be integrated. To start one, the economy must be strong. To start, the world economy must accumulate wealth to distribute it to the rest of society. To distribute it to the countries and communities that need dire help. Will it work? Yes, but it must be in a liberal environment. A liberal world. A liberal order. It may work conservatively, but they must distribute wealth liberally. This means it must be distributed to the right people, not just the elite and powerful.

As I mentioned, the key to the World's Golden Age is the middle class. They must focus on accumulating wealth for the middle class or risk turning the lower class into one. We must focus on it. We must aim for it. It will be hard and daunting, but in the end, we will all flourish crossing the finish line. No one will prevent us from proceeding, as the economic disruption will be identified in advance and resolved. We must accept that not everyone will cooperate on this daunting task. It will also be a task that is extremely hard and complex, and we cannot let anyone fall behind.

I must carry out this daunting task. After all, I have no choice because my ideology perceives me to pursue such a task. To pursue an objective so complicated that it needs to be focused to ensure success. That will change the upheaval of society. That it will be accomplished. That will be the manifestation of the world economy. That it will be God's work. God's reality.

So, let us go back to the daunting task at hand. First, we must understand that the goods and services distributed to and from such an economy must be effectively managed to ensure its integrity. Of course, it can be done, but we must know that it is the entire world about which we are talking. No one must be left behind. We must focus on the countries that are in dire shape. Ensure they are integrated and become part of this complex world economy. We must ensure that the task at hand is done correctly. This means that wealth must be redistributed to the right people. To the right community. To all members of society.

The World's Golden Age can be a reality. We must ensure that it does. We, the people, must work together to ensure that it will be a reality. We must not let ourselves down if it doesn't work tremendously, as it will be a disaster, requiring significant effort. That means we must focus on building jobs. We must focus on building wealth. We must focus on building roads, buildings, and bridges. We must concentrate, for such a task will be nearly impossible. To ensure the World's Golden Age, we need to ensure that economic growth is necessary. Economic growth will not be disrupted. We must ensure that everyone is working to ensure such economic growth.

China is one example of this. Their economic growth is so tremendous that everyone is surprised. India is next. India must be the next China. India must work hard to ensure economic growth.

As I said, no one must be left behind, for God himself will have tremendous arduous work to accomplish such a task. But unlike God, we can work together to carry out such tasks. We must work together. We must focus. We must face it. We will proclaim victory. We will do God's work. We will do God's task.

I am a luminary, and I will do such tasks. But I am alone. I am unfocused. I am not in the shape to do such a task. For I am not confident to do it. For I have no confidence at all. That is because it is a daunting task, and someone better than I can perform it. Someone who has knowledge and power. Someone whose leadership is tremendous. Someone who is tremendously focused. Someone who can do God's work. I am hoping he will read this book. For one day, he will accomplish true greatness as equal to God himself. He does not need to be a World conqueror, but he will unite the world. He will probably be equal to the next Messiah. God bless him and God bless us all.

I have seen greatness, but it will not be accomplished
If you think that I am only normal
For your love and confidence will ensure that

Let me be your guide.
Let me be your soul
For you conquered my soul
You conquered my world

Chapter 8
Integrated Economies

(The Unification of Nations)

Unified Nations. It is an integration that can only be accomplished using trading and business development. It is a privilege that must be done to prevent and limit war, to find peaceful ways to cooperate with other nations, to be a self-worth nation, to start, continue, and accomplish a goal, to prioritize that goal, to seek accomplishment, and lastly, to grow and prioritize the economy of such nations. Such accomplishment must be done with precision, collaboration, and prioritization. It must be done for the nations to integrate, grow, prosper, develop, and maintain their economy.

If such integrations exist, they must be done without any disruptions in either internal or external factors, such as trade tax, tariffs, embargoes, sanctions, and more. To do that, one must understand the nation's economic intricacies, trade barriers, industrial varieties, businesses, and government programs for companies and workers.

Another aspect that must be accomplished to integrate the

economies is trust and investments, also known as Foreign Direct Investment. Such priorities must be done to ensure the flow of money will not be disrupted by graft, corruption, and many other issues. Corruption is the worst entity that disrupts the flow of money, the economy, and businesses themselves. One must know that corruption is the worst entity that destroys a nation, and it is being used to influence money and power. Such an entity must be prioritized to ensure that the flow of money is stable.

Economic integration has been completed and is ongoing now. Many such integrations are being done, including those of the European Union, Association of Southeast Asian Nations (ASEAN), G-7, BRICS, the African Union, and many more. However, to accomplish my aim and ideology, global economic integration must be done and prioritized. Such an accomplishment is so complicated that it requires the cooperation of most countries. Such cooperation cannot be done right now due to the power and influence that countries prioritize. They all want to prove that they are better and more powerful, and they need to ensure that their power is influenced by the businesses and other nations that want to trade with them. The European Union, the United States, China, Japan, and other developed countries are exercising such power. That power needs to be extinguished. It must vanish. It must not be a priority. That is because nations that are still developing or emerging need them to ensure that their people will live happily and flourish.

The goal of this chapter is to ensure that every nation becomes a developed country within a few decades. Such a goal is challenging and daunting, but it needs to be prioritized because every nation must have enough consumers to achieve wealth. It will make other nations wealthy as well. It would make every nation wealthy if the consumers' needs were met, which is the goal I wanted. If that works, then the Global Renaissance will be accomplished. As I mentioned, it just needs five to six billion middle-class people to make such a phenomenon possible. But they need to be integrated to provide jobs, wealth, the proper flow of money, business growth, and economic growth. Such daunting tasks are challenging for every nation to cooperate on.

But we do not need every nation to cooperate. But many or most must do it. The Belt and Road Initiative has proven that. The West must answer what it can provide to counter China's program. The country is becoming more integrated into the global economy due to its wealth and influence. In contrast, the West is declining, except for the United States, because it lacks the business and economic impact that other countries seek. If that is the case in a decade or two, once China becomes the number one economy, it will be hard for the West, even the United States, to fight back economically. As I mentioned, the West, except for the United States, has an anomaly: it has not provided a technology capable of competing with or dominating the entire world. It needs that. It must be prioritized. It must be accomplished. We have a decade or two to do that because once China has enough money and power, its influence will not be stopped. Remember, it only took a generation for the country to become what it is right now. You must know that it can take any action to accomplish its economic priorities in the future. The United States will not be able to stop that. It can disrupt, but China will be unstoppable. We must find a solution to that anomaly.

We always have choices to be made
Those choices will make us or turn us into dust
Understand that choices are the key to the right path

I am a cheetah
I am the fastest in the world, but I do not have the stamina
You are a gazelle
You are not the fastest, but you do know what to do

Chapter 9
The Ultimate Unification

(A Must-Enduring Unity)

The unification of society can be done in many ways. It can be achieved by entities such as innovative technologies, including the internet, electricity, and many more. It can be achieved through influential mega-building projects such as the Pyramids of Egypt, the Acropolis of Athens, the Colossus of Rhodes, the Mausoleum at Halicarnassus, and the Taj Mahal of India. It can be triggered by phenomena such as war, pandemic, disaster, or religious influence, among others. It can also be done by human beings, such as Hitler, Napoleon, Genghis Khan, Caesar, and Alexander. These are the ones that unite a nation. These are the ones that join a specific society.

However, we are not referring to a nation or a specific society. We are talking about the entire world itself. Can it be done? Can it be influenced? Can we unite the world itself? Currently, the answer is no. We cannot unite the world in its current situation. But in the future, we can do it. The problem is that geopolitics will always be the barrier. It will always be the reason we are not united. There are currently two countries competing to dominate the world. Their power and influence are so immense. The first, the United States, has been influential since the end of World War II. The second one, China, has recently been promoted to contender status. These two countries have been influencing the world into two humongous alliances. One is with the United States, which is very influential. The second is China, which has a rapidly growing influence. Their economies are so huge. Their military is humongous. The only difference is that the first one has a power that is already at its peak. The second one is getting there fast.

However, of the two, only the United States can influence the entire world if done right. We have Hollywood, which influences entertainment, and Silicon Valley, which influences technologies and entrepreneurship. We have the media industry in New York that influences the whole country's journalism and media. These are the entities that influence the Americans. These are the entities that we are using to influence the world.

But I am not talking about influencing. We are talking about uniting the world. Yes, it is just an influence, but one day a leader can use these entities not only to influence the world but also to manipulate it. We must use it for World Unification. We must use it as an advantage. The American People can also be used to unite the world. They are the most innovative, most influential, and most resilient. They can adapt to changes easily. They can be phenomenal. The American people are incredibly confident. They are well-determined. They do not want to lose.

The problem right now is that the American people are divided. Someone must unite them. Someone must influence them. Someone must give them a direction to the right path. Of course, I will not be that person. I am just giving you some advice. I am just giving directions. We did it once. We united the Greatest American Generation -

the generation that survived the Great Depression. The same generation that fought in World War II is now aging. The same generation that provided directions to the next generations is now guiding them. These generations are old and dying. We must have something to replace them. The problem is that there is none now. We must create one. We must start a generation that can do that. It is probably too late right now to be part of that generation due to the ideologies that divide us. However, the next generation must be taught how to remain united despite differing ideologies. This generation must know how to value respect. This generation must value patriotism. The generation must understand the entire world itself.

The next generation will not only influence us but may also impact the entire world if they are taught correctly. They will if they have discipline. They will, if God is on their side, and I am sure He is.

Of course, to unite the world, a leader must be proclaimed. That leader must be one of the smartest, probably the smartest. He must be very disciplined. He must be very influential. He must speak as influentially as almost a Messiah or probably God Himself. He must be something. He must be a unifier for both Liberals and Conservatives. He needs to be because that is the only solution for this country right now. A solution that needs to be tackled. The solution that will unite the government itself is required.

And the most crucial part is the language itself. Yes, English. We must thank the British for that, for they spread it around the world. We must acknowledge the power and influence that brought the language to all corners of the world. Now, we must take advantage of that. The language is already influential and is the most spoken language in the world. Of course, Mandarin of China and the Hindi language of India are also influential, but they are concentrated in one area.

However, the United States cannot do the job alone. It needs help. It needs other countries or groups of countries to do the job.

And that Europe and East Asia will come to mind. Allies indeed, but they need to have the same quality, perception, and ideology as the United States. However, that is going to be complex and daunting because most European economies are stagnant. Japan and South Korea are performing well, but they also have their issues. But the ones that would really follow are the developing countries, for they want a sense of direction. They want to know where their economy will be going. What will be their finish line? They need a significantly influential country that can provide them with guidance. They will indeed need the United States. They may go to China. But once we can influence them, then we can be chosen over China. Just like the Internet, which we helped release, we were the envy of many. We need to be envious again by providing innovative technology or technologies. We can influence them using these technologies. That is because we are going to make a lot of wealth. They will accumulate a great deal of wealth. We must provide it to our allies and developing countries who want it.

My country is small, but it is not all about Geography
It is all about the People
It is all about the Society
For we are small, but we are Great

I am a Blessing in Disguise
I may be invisible to you
But in the end, I will be your light

I am your Black Hole
You will not escape my Event Horizon
You will not be spaghettified
But you will not escape my love

Chapter 10
Five Billion Middle Class And Upper Class

(Yes, it can be done!)

I have been under mounting pressure on how to perceive this .... Unfathomable, ridiculous, and surmounting idea of the five billion middle class being the solution to the impoverished world. Such an absurd idea will not be taken lightly by countries, as it is simply too ridiculous to be considered. However, I have to provide the solution, as it is my only choice. It is my only doctrine. It is my enlightenment.

There are currently approximately four billion middle-income and upper-income individuals worldwide.[5] We are improving and becoming more confident that we will achieve more. This is what my book is all about. To guide countries around the world on how to address this phenomenon and achieve a breakthrough. It is a process that will probably take a decade or two. We need one billion more middle-class individuals, primarily in India and other major economies around the world. India is poised to become the most invested country in the world, thanks to its 1.4 billion people who are increasingly confident and forward-thinking. India can achieve such requirements with a meticulous, productive, confident, and almost perfect system. A system that can be provided to society. A system that will work well for investments, productivity, commerce, and manufacturing. What India needs to do is follow the path China has taken. This is not the path of communist doctrine, but rather the path of a liberal economic system intertwined with a socialist economic system with Chinese characteristics. They have to do it because China managed to do it in just two decades. Now, they have six hundred million middle-class people eager to engage with the world economy. India has to be like that, too. They have to provide another six hundred million middle-class individuals, which I believe is doable, but the complexity is just too great. India has a democratic system with complexities that need to be addressed, provided, and implemented. There will be many reforms that will help with the progress of Indian society. There will be a lot of progress that needs to be made. Yet, we have to prioritize the Indian people, for Democracy is all about a free society. Though China is a free society, its leadership is not. But it works because for them the job is easy. It is not that easy, but it is nearly unstoppable to implement.

India can do it. It does not need an authoritarian government, as was done by Taiwan, South Korea, Japan, and Singapore. Once those countries' economies met the standards of the developed world, they transitioned to a democratic system. Investments poured in. And the people live in a better environment. As I said, India does not need a dictatorship. It is just too complicated. The people still influence Mohandas Gandhi. They believe in a free world where conflict can occur without any gunfire. Gandhi influenced them to do it. His modesty, simplicity, and vision will always be the right path for his people. His revolutionary work is a model for all. His people will do challenging work. They will be unstoppable. They will not stop, for that is what they will do to their economy. That is what they will do to the Indian society. A growth rate of 6% to 9% for the economy is sufficient. But the most feasible way is a growth of 10% to 12%. Why? That is because, done that way, entrepreneurship, labor, and investments will be gradual. Of course, it will provide more opportunities for middle-class and upper-class individuals. It will give more consumers. It will help the country more. It will be another great miracle. It will be unstoppable. I suppose it could accommodate around four hundred million middle-class and upper-class individuals.

There are other countries that can provide more opportunities for the middle class and upper class. Those countries must be major economies, such as Brazil, Russia, and Indonesia. Those countries have more than a hundred million people each. With their population comes the consumers. The middle class and upper class will continue to grow, leading to more demanding consumption. An additional 200 to 300 million people in these groups are needed for the prescribed countries to grow.

As I mentioned, all countries listed are developing countries or emerging economies. Precisely, I mentioned BRICS, which are the main driving force behind the world's economic growth. They will make more consumers. They will create more middle-class and upper-class. They will help in integrating the world.

Of course, the rest of the world can help us. They just have to provide more consumers. They just have to provide more middle-class and upper-class. There are countries that I did not list but may work as well such as Pakistan, Bangladesh, Vietnam, the Philippines, Thailand, Mexico, Malaysia, Saudi Arabia, United Arab Emirates, Egypt, Iraq, Peru, Chile, Colombia, Qatar, Kuwait, Nigeria, South

Africa, Kenya, Oman, Mongolia, Azerbaijan, Kyrgyzstan, Turkmenistan, Uzbekistan, Tajikistan, Kazakhstan and many more. These are developing countries, but the opportunities they create are insurmountable for them. Let us say they can provide at least three hundred million middle-class and upper-class individuals with support.

Remember, Economic Integration is the key to providing this solution. We have to be precise about that. We have to invest in them. China has done so with the Belt and Road Initiative. Everybody else has to work together to do it. IMF and World Bank can help, but we need more alternatives. Once complete, the Global Renaissance or the World's Golden Age will be a part of our civilization. Overall, the calculation indicates that there are nine hundred million to one billion middle-class and upper-class individuals. That will bring the total to four billion.

Such calculations are formidable. Though it may work, something must be done. My previous chapter, "Unifications of ....", will serve as the watermark or similar integrated designs. As I mentioned, Technologies are the keys to unification, imagination, broadened minds, and clear ideas. But just like the Beatles, Tomorrow We Never Know, but it is always a Day in the Life. It is always a complicated question that needs a gradual result. Mine are just a guide. It is up to you to make it a dream and turn it into reality. For we are one. We will be unstoppable.

Once we provide and develop such Middle-Class and Upper-Class Services, we must know the results. And what will it be? First, more jobs will be created for businesses that need them. Second, more communities will be established, including business establishments like offices, factories, and other facilities. Subsequently, more housing and living spaces will be created. Third, more Government Revenue will be generated due to the increase or explosion of the middle-class and upper-class. That means more taxpayers. As more revenues are made, more expenditures are being made as well. That makes the budget more formidable because it will shorten the budget deficit or create a budget surplus. With that, the Government can find ways to tackle projects such as infrastructure or directly fight poverty.

For real, I would rather be a child than be an adult who acts like a child

But sometimes I want to be a child to forget that I am an adult

Chapter 11
And The Rest Will Rise

(They will make it. They will find Enlightenment)

We make choices. We have the freedom to do so. For we are formidable. We have indulgence to ensure such, for we are human beings. We are God's choice to dominate this world. We live a life to the fullest. We become tremendous. For the choices we made, we were fulfilled. For we become more than humans. We are now almost immortal.

However, the impoverished make no choices. They are in a world with no hope. They are in a world of dead ends. They think that society has neglected them. They thought society did not care at all.

This is it. The most important chapter. The most significant part of this book. To ensure peace and tranquility, we must provide a solution to people experiencing poverty and address other related issues. To all of us.

We can create a formidable middle class, but we have to ensure that the rest will live not in infamy but in harmony. We have to help the rest.

So, what ideas and opinions can I provide to help? There are some, but I am hoping others will offer more. I am a lone person with ideas that have limits, but the collaboration of many can lead to miraculous, unending, and insurmountable ideas. But we have to ensure that the ideas we provide are not too extreme, as the result may lead to disaster. It may lead the community to hell. It may lead society into condemnation. We have to be moderate. We have to be meticulous.

The first idea I can suggest is to provide them with jobs. A lot of jobs. Of course, to create jobs, we have to establish a lot of businesses. The government can give as well, but the key is private companies. They will make a change. They will create millions of jobs. Of course, to start and flourish a business, our economy must be strong. As the economy improves, more and more businesses are being established. That means more jobs. That means more opportunities. And the economy will continue to improve until it peaks. The flow of money will be easy, and people will flourish. Remember, jobs are the first choice to ensure a constant flow of cash through income. To ensure wealth is distributed well.

The second idea I can suggest is a basic income. Yes, it needs to be done. We have no choice. As I mentioned before, basic income does not need to be big. It has to be mediocre. It must be to cover crucial expenses such as food and shelter. You may include important bills in that, too, but it is not necessary. But someday it may. Basic Income is probably more critical to the poor class than the middle class, but we have to ensure everyone benefits from it. But let us make sure that the rich will have an option to ignore it. The middle class will probably neglect it as well, so they can make their own choices. But in the end, we have to ensure that people experiencing poverty are given the utmost importance. We have to prioritize them, but we must also ensure they don't get lazy and that they play a part in the community.

The third idea I can suggest is to do micro-lending to people with low incomes. It was an idea made famous by the Nobel Laureate Muhammad Yunus. He proposed that poor people and impoverished business entrepreneurs have a chance to get money when no other banks can provide it. Such moves are critical because most of these people have little or no money at all to pay back. His idea and strategy worked well and proved to be the right idea to serve people experiencing poverty. Muhammad Yunus' idea is to create a poverty-free Bangladesh and ultimately a poverty-free world. For more of his ideas, read his book called "Banker to the Poor."

So how can we cover these expenses in the budget? I know it has been going on for a while, but it will still work. And I will say, let's tax the wealthy. Yes, we must tax them. Not tremendously, but just moderately. We know that the rich may provide wealth to the poor with charities and donations. The problem is that not all of them will do it. Some will try to get away from it. Some will sway it. We need a way to tax them. Yes, they can have mega-yachts and private jets. They can have mansions and more mansions. They can have more cars. Yet, what fulfillment can it bring to society? I do not mean to say that I have to stop them from doing it. They need to know that there are people out there who need help.

Remember, we cannot have another French Revolution. We cannot do that. We cannot have an inequality that is so unprovoked that it cannot be allowed. Such moves are a disaster waiting to happen. We must uphold inequality to almost "equality." We cannot blame the rich for being rich. This is what the system of Capitalism is all about. Rich must be made, and competition must be allowed. That made the United States one of the most competitive economies in the world. We are confident that our competitive society will bring about economic upheaval, ensuring the status as the biggest economy in the world for more than one hundred years.

But to be the biggest economy comes with small sacrifices and dignity. Such moves are needed. We did it in the 1990s, and our economy became the envy of the world. Budget surplus was once seen as a myth. It only happened five times in the history of the United States. The last time it happened was in the year 2001, when the Tax Reform had not been enacted yet. However, if the Government budget is allowed to balance unopposed, then it will happen. But it must be done with the newly introduced technologies. It must be done with united parties. The country will improve. Businesses will be made, and jobs will be created. The people and the government will make more income. Economic growth will be enduring not only in the United States but all over the world. We have to be confident and imaginative that such an occurrence will happen again, just like the 1990s.

The fourth and final idea I can suggest will also help people experiencing poverty. It is an absurd idea, for it will create a diabolical system that proves to be socialistic. Such an idea cannot be bound to a Democratic society, for it proves to be extreme. It proves to be an anti-capitalist. Yet I must provide an idea, for it is the only choice. For I thought it was the only worthy answer. For I thought this was the humane way.

The idea is to move them in a Colossal Way.

Why? That is because it has already been proven by the country of China. Yes, they are a communist country, yet they are capitalistic as well. They combined the two to ensure a significant upheaval in the economy and to bring the people together. For the mind of a person who is in the West, it proves to be extreme. But for me, it is not. For what I say will not prove extremism. That is because the West has done it before, but only in a minimal way. China did it in a colossal way. It was so colossal that extreme poverty was finally eliminated in February 2021.

So, what did they do? As I mentioned, they just moved them. They moved millions of poor people from an area where the environment is scarce. The government moved them to an area where resources are abundant. Their homes become better with free utilities. The government did not take their farmland. The only problem is that it's a bit further than their previous home, of course. Some people were provided with new jobs. However, many of them know that this is the opportunity. This is the time. They must be hasty, for the opportunity will disappear. But it will not disappear. It will keep coming. Some people experiencing poverty are already planning to move to the cities for more opportunities there. The government is preparing for more affordable housing. Millions of them. They have already eliminated extreme poverty. Their next goal is to accumulate more of the middle class. And I mean the entire population.

This is the Chinese Dream. We do not need to follow them. We just have to find a way to help the impoverished. Not in a colossal way but just in a moderate way. For we are Democratic, we must be moderate on all things unless it is crucial like another Recession or War. It must be moderate because the budget will be moderate as well. We do not have the money capacity or capability of the Chinese government. Yet, as the people, we can do such things. We can provide free housing or build ones.

We can work with the government to ensure one. The key is to volunteer to ensure poverty alleviation. The government cannot do it alone. They need people to help. That is because the Chinese government will implement such measures using their system. And we cannot follow their system. It is just extreme compared to the rest of the world. The rest of the world has to do it on our terms, our morals, our dignity, and our obligation.

The Chinese power and influence are getting more influential in this world. Many developing countries are engaged with their economic power, trading, and influence. The Belt and Road Initiative was the most significant influence they had in the world. These countries might use their system to alleviate poverty. The question is, will it be feasible? Will it be manageable? Will it work? That is a question that will not be answered for years to come. We will have to wait. It may combat poverty, but will it do so with a budget that the rest of the world cannot afford? Let us see.

In Economics, there is no such thing as a Free Lunch
I hope a Right Society has to be made to ensure one

Myths and Legends
They are just Dreams and Imaginations
But you are a Myth, you are a Legend
You are Real
You are the Greatest

Chapter 12
My True Political Ideology

(The Real Lifetime Choice)

Democrats... Republicans... bunch of formidable foes... cooperation is always a disaster, yet formidable as it seems, united and wonders are made like God's creations. As if a divine entity had just been created. Yet that is impossible now because these foes lack a sense of unity. Have no sense of ideological compromise. Have no reason at all.

Once, they had no boundaries to disrupt their unified ideas, but now those boundaries have all but vanished. Once, they were intertwined in their beliefs; now it is just a thing of the past. All are like broken mirrors. All are dead ends.

So, what to do now? What must be done? What are the choices that need to be made? For our freedom becomes corrupted. For the country is now full of holes that cannot be patched. For these two parties will not unite. For these two parties are just ideologically savvy.

Kevin McCarthy – Former Speaker of the House and a moderate Republican. There is something in him that I cannot fathom. There is something in him I cannot relate to. For it seems the speaker knows how to fight. The speaker appears logically a uniter. It seems he is our last hope. It seems both parties can tolerate him. The people can accept him.

The only problem is the Extremes are just too much. They thought they were formidable. They thought their ideology came first before the people. They believe they are God's servants. But they are living in an illusion. They are living looking in a broken mirror. They are living in a world of rodents. They thought the Speaker was wrong and believed they were right.

The problem is that the Speaker is constantly threatened because he does not like to commit to the ideology of the group. For he believes in moderation. For he is a uniter, not a disaster.

Ultimately, he will be a luminary. He will carry the light. He will be the guide to the next phase of generation. I am hoping he will unite the two parties. He will unite the Government. He may not become the next President, but he will have my vote. He will have my confidence. He will have God's Given Creation, and that is the United States of America... And the rest of the World.

But it's awful

I know it's awful....

Definitely... Tremendously

Did it change you? Yes. Did it disrupt your faith and will? Yes. Did it make you better? PRECISELY. It has to be changed like that. The blood that you sweat. The hard work you persevere. The tremendous, awful, and unimaginable actions they did to you. But do not declare that you lost. Do not declare that it's over. You are bound to this environment. You are bound to be a great leader. Just to let you know, in life, you have to make choices. And the right choice is a must and must be thought through. The intimidation, distraction, and humiliation will not abrupt or disrupt you. It has to make you better. It has to make you a better person. That's because you can't be on the upside all the time. You have to learn how to be on the downside.

It's not what they did to you that destroyed you. However, youfollowingxt action is the most crucial decision you must make. And that is what you need to understand. Of course, life must go on. But it must be in the right direction. And those directions will guide you to recovery and help you become the person you want to be.

In the end, the audience, representing society, will provide you with the final verdict. It could be anyone. It could be your wife. It could be your children. It could be your friends. It could be just a beggar in the street that you always pass by. And it has to be all of them who have to say that you are a good man. A model for the world. A model for all eternity. For they will remember you today. But the most important thing is they will all remember you for a thousand more years. That's because you are not just a good man. You are the man!

You have to let it out.
Do not let it consume you.
Or else you will be an explosion
You will not be the same after that
But they will still give you another chance
Just don't explode again
Because you will always be the same after all

In the end, we are all the same
We all die
But our ideas and legacies will live forever

Chapter 13
Russia

(You are not the Great USSR Anymore, but you are still Great)

It hurts a lot. It hurts a lot. The Great Russia or USSR turned the tide of their enemies in World War I and World War II, and even the Great Napoleon Bonaparte lost not in war but in peace. It devastated the country so much that they changed the world to end in a catastrophic insult. It is a disaster they cannot fathom. They do not understand why they lost in such devastation, for they are a Great Empire. An Empire that is feared and respected by the rest of the world. The loss was due to the Berlin Wall Catastrophe, which, for the West, is a Great Victory. Yes, it was a devastation so tremendous that even people today cannot understand why a country so powerful was lost in that event. Is it because they made a mistake? Is it because the West outsmarted them? Or is it just misfortune?

All I can say is you do not want to go back to the Great USSR anymore. You are now Russia. But you are still a country of greatness —a country with smart and proud people. But you are not an imperialist country anymore. For imperialism is now obsolete. For everything is about the economy now. It is not about war and colonialism. These are obsolete now anyway. This is a history that is shameful and will not be repeated by the West. It is about society now, and that needs to be kept in mind. Society fuels the economy, institutions, innovations, and most of all, peace and stability. Society is something that must be considered, and Russian society is one of the greatest. Your society is so advanced that you put the first satellite and man in space. You stop the likes of the Great German Military of the Nazis and the Great French Army of Napoleon. And they are not just great military. They are the most significant military forces in the history of humanity. And God Damn! You did stop them with your most excellent strategy, of course. And that is, you made them underestimate you. You did lose a lot of men in the war, but you never quit. But it's ok. It's about the result, of course. And that is to win the war. Don't forget. The whole world loves you for that. Both wars of course. Because you saved them, you saved the world as a whole, of course. Napoleon is a good man, of course. But he can't rule the world. No one can. Those are the rules, and they will always be the rules. Hitler, on the other hand......

The best thing for Russia is not to think of itself as the Great USSR, but to consider that the world is becoming more integrated. It's getting intertwined. It's getting better.

Change of leadership is one thing, but change of engagement with the rest of the world is a must. A thing in the past will not occur again in the future. You have to remind yourself that the past is history you can follow, but it will not be repeated in your current position. But you still have a society that can shape who you are now. The future lies in these young ones who want change; you have to invite them to be part of that transition. The formulation of change doesn't continue with the leadership; it has to start with the young ones. The energetic ones who will do everything they can for what they believe in are the ones they want to absolve. They want a change, and they will make a lot of sacrifices. They did it before, and they will do it again. A change is a change; they have to believe in it.

For society will die without progress. For society will not be the same if everything is the same. It has to change, and people have to change. For that, humanity needs to know that we have a world in need, and Russia needs help. They can do it on their own, but they need someone who can help and trust them.

The country, of course, will be the country with a lot of future. Of course, the young ones believe in it. They know it's coming, and they will sacrifice themselves to ensure such a thing will occur. All they need is for someone to believe in. Of course, it's not this book but a leader who can help them all. He has to guide them to their maturity and dignity from their innocence.

This country still has a lot of potential. Of course, they need to build expressways, rails, bullet trains, maglev trains, and many more. They do have a big population, but they need to grow more, and demographics need to be younger and bolder. In the end, they will succeed. They have to achieve, and they know that.

There is something about the land to the east of the Ural Mountains in the lands of Siberia. It seems barren right now, but a long time ago, like millions of years ago, it wasn't barren; instead, lava and some volcanic openings covered a large area. Does it mean something? Probably. Usually, land formation follows normal processes. It occurs naturally that we have earthquakes and volcanoes. But it seems that Siberia at that time was just full of lava, and it has been doing it for millions of years. It is called the Siberian Trap. And of course, why such a massive volcanic event? Is it a danger to the territory or an opportunity? We have to assess that. I know there is something on the underground that makes it unusual. Something that can provide heat to the whole Siberian region. It will, of course, melt the snow and provide the ideal temperature for the habitats and humans who will use the place. Of course, there is no such technology to assess that. But it's a technology that we can enumerate in the future. It's probably geothermal, but it needs to heat the environment, which I think will be helpful. And that environment is so big that it will be the most significant opportunity not only for the Russians but for the rest of the world. I mean, Siberia is the most critical land mass in the world. It is just too cold, but with the right geothermal technology, it can provide just the proper temperatures. Of course, it will create cities, farmlands larger than anything else, and numerous opportunities for the Russians. But we have to work to ensure that geothermal energy exists. Of course, it will be the young ones of Russia, and they will do everything to make it work out. So, the Russians need that technology, and it will work. The geothermal technology I am referring to will heat the atmosphere and the entire region of Siberia. It will work out, but we need to ensure that the underground, which includes the crust and likely the mantle, can provide it. Because I believe the mantle in that region is not metallic, which is the standard form of the mantle, but it's probably soft and probably dangerous if not properly assessed. But the opportunity has more gain than the disastrous outcome. We need to check the crust and the mantle to ensure the safety of all of Siberia, which, of course, is almost impossible because it is just too big.

But Russia has to do something responsibly. They have to cool off the Arctic again to ensure Climate Change denialism. Can they do that? I believe they can. They need to work on the assessment to achieve that. It's their territory anyway, and they will do everything to put that ice back again. Of course, no technology can provide that right now, but I believe they can, as it's also a reward they can win and absorb. But Climate Change is real. It could be artificial or natural. But it has to be tackled, and freezing the Arctic is the only way to address it. The Antarctic is also melting, but anyone can re-ice it. Is it the only way to tackle Climate Change? No, there are more.

But I am just talking about Russia, and that will be another topic for the day. But hey! It could be another chapter. Nah... I had it with that—too much politics. I hate politics, and I don't want to tackle that. But hey! Climate Change can be natural!

The only thing you can do for Russia and the Russian people is to intertwine with the world, not through your natural resources, but with your great minds that can achieve amazing innovations and provide ideas that will continue the world's progress. The oligarchs might be powerful, but they are people who can be changed. They will change with the right mindset and idealism. We have to make them contribute more to the Russian Society by sharing their wealth, of course. And I believe many of them are probably doing it. But you do not need to think like the Great USSR. You have to feel like Russia with a Free Society that can do wonders. To prove it, you can do it many times in history. You can do it. As I mentioned, a flamboyant society is probably a must, and Russia is one of them. They need to think and integrate like one because they can, and they will continue to do so. They need to be confident, reread their history, and remember that they are always great and capable of doing wonders. I have another proposal for Russia, but it is in the chapter called the Sphere Theory. That one is a game-changer. And it will accommodate Russia more than others.

It is now and always Great Russia, anyway...

God may have a limit.
But to us, we always know he is unlimited
But as I said again, he has a Limit

Chapter 14
An Almost Inevitable War

(It Will Always Stop, But It Seems Unstoppable)

Despicable. This is how the chapter will be. Unfathomable. This is how the chapter will end, as it contains violent content. It will be annihilating. It will have an invocable desire for us, but it will be immeasurable for the result will remain almost inevitable. For war will always be unpredictable. It will always lead to neglect and disdain for our morals, affecting our dignity. It will change lives and, of course, end lives.

We have been in this situation since the beginning of humanity. Civilizations rise and fall due to war. The greatest empires that were created and annihilated due to war. We become what we are now because the outcome of the war will alter our perception and vision of the world.

We are now in the midst of that war. Our world would change if that war were to happen. Our lives will never be the same. So as the society. So as the civilization. For a great war may come. It will probably be the last war. The final war. The war that will end us all.

Taiwan and China have been in an ongoing feud for more than a hundred years. They have been in a state of separable, unrecognized, and unrelenting relationship ever since. They try to be unified again, but there are always entities and forces that keep them from doing that. They have been in this situation since the Kuomintang was ousted from Mainland China by the Communists in 1949[6]. They have no choice but to move to the island of Taiwan. The last one remained. The only choice they have. I know it is insulting, but that is what you get when undermining the enemy. They have been stuck on that island for a long time, and they do not know how to get back to the Mainland.

With the help of the West and other countries, Taiwan is now one of the technologically powerful countries in the world. Their companies are so advanced that they are the envy of the computer hardware industry. They have many good brands and influential technological companies, with one even managing to top them all – TSMC. The company is so advanced that no country, not even the United States, can match its technology. They developed computer chips, but ones of very advanced design. While the rest of the world is still developing three-nanometer and seven-nanometer chips, they have already moved to two-nanometer chips and continue to innovate.

China, on the other hand, is trying to develop new computer chips, but someone is on the way. The United States, once it found a weakness in the country, took advantage of that weakness and decided to sanction China from getting not only advanced computer chips but also the technologies that can make them. And that is Lithography. The most advanced machine that humans have ever made. It is so advanced that it can produce chips so small they are nearly impossible for other machines to create. It is also one of the most expensive in the world, probably the most costly that was made. It is made only by a company named ASML in the Netherlands. They are the only ones who can make it, and the United States will not allow them to sell their products to China.[7]

That is the disaster of sanctions. It is something that must be stopped. That is because I believe in Free and Open Trade. I do not think so of Protectionism. This led to an escalation of the Great Depression in the 1930s, with countries in Europe and the United States adopting protectionism to safeguard their economies. That led to the inevitable destruction of their economies and exacerbated the Great Depression.

However, China is relentless. They have the money and will to make chip technologies. They are trying to make their chips more advanced and smaller. But they know it will take years to cope with the current technology, as the advancement and innovation of chip technology have taken decades since the 1980s. However, once China finally produces these chips, it will be a disaster for chip-making companies, as their products will face unfair competition due to their low cost (or not?). That is the reason for the sanctions. But there is another way. It is called subsidies. The World Trade Organization has been provoking it in Western countries. Subsidies are funds provided by the Government or a public body to ensure that companies' products are affordable or competitive. That will work well for the chip industry. That will work well for all sectors. But the Government has to prioritize the products that will influence society, and chips are one of them. That is because the chip is being used everywhere, such as in vehicles, computers, home products, industrial products, and many more. It is now the most essential product that society can benefit from.

Another issue that Taiwan and China confront is the geographic importance of the island. The island has a strategic benefit. It is near the allied countries of the United States. If invaded, it will divide the alliance. For example, Japan and South Korea, to the North, will have a tough time integrating with the Southern Alliance of the Philippines, Thailand, Malaysia, and Singapore, or most ASEAN alliances. That is because Taiwan will be in the center, blocking their integration. That happened during World War II when both Japan and the United States invaded the Philippines. The strategic importance of the country led to the invasion of Southeast Asia using a mediocre military strategy. That is because the countries are now cut off from military supplies and ships from other allied nations.

However, that is not the real reason China will invade Taiwan. It is not about technology or territorial strategy. It is always about the One China Policy. The policy integrated the mainland into the island. It is about a plan to unite the people. It is a strategy to stop both from interfering. However, Taiwan is a democratic government and a democratic society. They will not change, and they fear change. They are afraid of China's dominant power. They are so scared of Socialism. They are fearful of losing their daily lives. Life in China is different from life in Taiwan. The Media is controlled by the Internet and many more. That will relent Taiwan to unification with the Mainland.

Yes, it isn't very easy. Taiwan and China have conflicting systems. That will be a problem for Taiwan because it is small, and its population is smaller than China's. China can easily invade Taiwan and will probably complete the invasion in one or two days. It will be fast, but it will be bloody. So, what China will probably do is make an amphibious force as large as, or even larger than, the Normandy Invasion. Remember, the biggest amphibious invasion was the Normandy Invasion. China will do that. They have no choice. The invasion must be completed in under a week. Not to win but for stability. To make everything normal, the next week or two after that.

The problem is not the invasion. The problem is the aftereffects of the invasion. The United States will be there to try to stop the invasion. However, with one carrier group (a group of US Navy ships with an aircraft carrier as its central core) protecting and securing the island, they have no chance to stop the invasion. China will win easily. They already have the biggest Navy and Army in the world. Soon, they will have the biggest Air Force. Their Air Force is already advanced with the J-20 Stealth Fighter as their primary model stealth fighter. And they will make a lot of them. The United States knows it, too. They know air superiority is a necessary priority and cannot be dominated or threatened by another country. They must win air superiority, for that is how they win in every war except the Vietnam War and the War in Afghanistan. But those wars are not conventional. They are more of a guerrilla war. The war against China will be traditional.

Once the invasion of Taiwan is complete, it will be too late for the United States and its allies to provide reinforcement. It will take days or, more likely, weeks. The only possible solution is to liberate the country. And that will take a lot of time. It will be a war of attrition. It will be an exhaustive war. A war that will probably take months or years to end. That is because China is a manufacturing power. The United States and its allies are no longer merely capable. They used to during World War II, but this time is different. Remember, this is the War of Attrition. More and more supplies are needed. And China has the power to do it.

The United States's only advantage is its allies. The United States and NATO are formidable. They have the best military in the world. The only issue is their bureaucracy. However, that problem can easily be tackled once the war starts. They will be more united, and they will not stop.

Remember, China can recruit a hundred million soldiers and other military personnel. NATO and its allies need to counter that.

They need to tackle China's military manufacturing power. Geography is another problem, as is logistics. It will be hard.

However, we need to define terms such as 'promiscuous'. For it is indeed just fiction, for it will not materialize. If China had invaded Taiwan, it would have done so a long time ago. It would have done it after the communists won the civil war. It could have been done after they won the Korean War. That is because their soldiers at that time were highly experienced and battle-hardened, making them virtually unstoppable. The enemies at that time were weaker, but China did not launch an invasion. Taiwan has a good relationship with China, both commercially and economically. Recently, Taiwan has been providing its technological prowess to China with no guilt. China is providing economic incentives that help Taiwan become a more technological power. But the recent conflict has diluted it. The people of China and Taiwan do not want reunification, but they want to collaborate and work together to ensure an ongoing relationship. The leadership of the two countries simply will not foster a better relationship.

However, the most critical aspect of the Taiwan-China conflict is that there is no actual conflict. China does not need Taiwan at all. Yes, Taiwan has many technological companies, but China has more and better companies. As I said, Taiwan's main technological product line is computer hardware. China's main product lines are so diversified and advanced that, among all countries, only the United States is more competitive. There is no conflict with people or society. Both societies collaborate with no issues or disputes.

So, war with China will not be inevitable. It will not happen at all. Although the leader of China, Xi Jinping, is considering it, he is aware of the risks and consequences if such an invasion were to happen. He is not a tyrant, but he knows he has to tackle it. The policy that has materialized for him and the previous leaders is to assess and investigate whether such an invasion will be necessary. For he knows, and the previous leaders know, that such an occurrence will be complicated and disastrous for China itself and, of course, the rest of the world.

And they know such an occurrence will be unfathomable.

The other side knows it, too.

Remember, once the War starts, it will be hard to stop it
It may be quick, or it may take years
But war is war
Everybody will be affected
And some lives will end

War will unite the People.
But peace as well
The problem is not the War
It is the Fear
People will be United for that
But so, as Love and Vision

Chapter 15
The War Of History

(It Never Ends, but It Has To. It is a Must.

"Jerusalem"
"Jerusalem"
"Jerusalem"

A city that works with a clandestine organization. Its history dissolves as different civilizations and empires have held it as a trophy. Holding it as a reward. Holding it as the center of their world. It becomes influential to them in their religion, culture, and most importantly, their "History."

The history of this city has been formidable. It has become since the Ancient age. Since the Sumerians, the so-called first civilization. The Jews were their descendants and a part of their culture. Abraham, the Father of Jewish Civilization, came from the Sumerian city of Ur. He then went to Canaan, which is now the land of Israel. His descendants multiplied but were enslaved in Egypt. Liberated by Moses, they returned, reconquered Canaan, and lived there. They conquered Jerusalem with King Saul. King David and King Solomon eventually made Israel a great nation, and during their reign, it became the Golden Age for the Jews.

However, too many civilizations and empires are just fond of the nation. It was conquered by just too many – Babylonians, Egyptians, Persians, Greeks (Alexander the Great), Romans, Byzantines, Arabs (Muhammad), Crusaders, Mongols, Mamluks, Ottomans, and the British.

During the British rule, too much chaos and disruptions were caused by both Jews and Palestinians. It escalated during the Rise of the Third Reich in the 1930s when more and more Jews immigrated to Israel (It was named Palestine during British rule). After World War II, more and more Jews immigrated to Palestine. Eventually, more chaos and disruptions occurred, escalating to a full war. The United Nations intervened by providing a plan to give lands or states to both Israel and Palestine and make Jerusalem a U.N.-administered city. That did not dissolve well. That is because Israel backed the plan, but the Arab Nations rejected it. Eventually, the war was won by the Jews, who established the "State of Israel." More skirmishes and wars came later, with no victory for the Arabs. Israel remains formidable to this day.

Then let us go with the Muslims, the next permanent occupant. The occupant lasted longer than anybody in Israel except for the Jews. They too have a history. They were peaceful, friendly, traders, and most of all, they accepted their Jewish friends even though they were the occupiers. They fought numerous battles against many enemies. And they live for more than a thousand years. The First Crusaders defeated them, and the Ottomans and other forces subsequently occupied them. But here they are, alive and well. They are strong, resilient, and seem to be the actual occupiers. For the Jews were gone, and only a few remained.

So why call it "A War of History"? Remember, wars and skirmishes have been ongoing since ancient times. The influence of the state and the city of Jerusalem is colossal on many religions and civilizations. These entities believe that the state and the city form a diamond of responsibility. A trophy is worth it as a priceless artifact. It gives them a mandate of significant accomplishment. A valid reward for those who believe this is the center of the world. The center of history. The center of all civilizations. The divine ones are in its history – Abraham, Moses, Jesus Christ, and Muhammad. This makes the place special. This makes the place formidable. This makes it the land that God chose. For it is the center of the Bible for both the Jewish and Christian worlds. The Book of the Quran made it special as well.

The real question is how to stop this fight between the Palestinians and the Jews.

There is no formal answer to that. But there is a possibility. The country of the United Arab Emirates, specifically Dubai, started it. Another city in the UAE, Abu Dhabi, along with other Arab states such as Egypt, Qatar, and Saudi Arabia, has followed suit. They decided not to empower their military. They decided to enrich their economy. If done correctly and carefully, these states will influence others, such as Kuwait, Jordan, Iraq, Oman, and many more. With that, they will influence Israel that military empowerment is not the right course anymore. It is about the people and enriching their lives. Make their economy better and stronger. Make them a desirable place to live. With that in mind, the paranoia that is always in the mindset of Israel will be extinguished. For that, the people of Israel will do that too. So, as the Palestinians are on their governed land. For they need it too. And their militant mindset will also be extinguished. This is the true key. The true answer to everything. It is not about Israel. It is not about Jerusalem. It is about the people.

It hurts that the Palestinians were made this way. It hurts that the majority of the Jews know, and they cannot do anything about it. It hurts the world, cries for an injustice that can easily be justified. It hurts because no justice was made, for history will never be history. Still, it is a religion and discrimination that results in unfairness, unrelenting hate, and, most of all, a never-ending war in history. The result of a dimension of not one-sided but too many sides forming into one, yet the occupant cannot abolish his doctrine. Does it start with the people? Yes. It begins with a continuity of history. It starts with the world working together, for we live not as occupiers but as a sanctuary. Palestine will always be a sanctuary for the Palestinians. Israel will always be a sanctuary for the Israelites or the Jews. They just have to work together to stop the hate. They did decades ago, but the Jewish Leader was assassinated by his fellow Jews. And that destroyed a goal. A solution for a continuous relationship. A continuous faith. A new history that will continue for the next thousand years.

Can it be done again? Yes. We just need a Leader and a Society that works as one.

Currently, the people's mindset is primarily influenced by religion. It is the history of transforming conflicts, religions, and power into economic salvation and integrity. For the intertwined Arab and Jewish states, it is possible. They just need to believe. For I believe in it. I think this is the only choice. You must remember that the past is just an illusion; the present is what we see in our soul, and the future is what we came for in this world. Let us shape the future with a new diamond of responsibility. For we are humans. We are human.

After centuries of wars, hate, and envy
We, as humans, can still change
We just have to believe
For how hard we are as a rock, there is still softness
that can make us think and feel like humans

Sometimes you forget the role of the Almighty
But when you need him, he will always be there
You just have to open up to him, and he will open up to you

Your gift will not be gratified if you live in dormancy
You have to find a way to open it
And once they knew you would live in enlightenment

Chapter 16
The Fragmentation Of Other Technologies

While software technology, information technology, the Internet of Things, Machine Learning Technologies, and Artificial Intelligence are advancing at a rapid pace, a significant transformation is occurring. Others are being bypassed. Others are being ignored. Others are being put out of place. Others are nonexistent anymore. What is going on? Why are these technologies not that popular anymore? There used to be a lot. It used to be the norm. It used to be a culture. It used to be the way we are, the way we were, and the way we will be. For these technologies were everywhere. These technologies are now gone. Fragmented is the right word. Remember, other technologies out there need to be recognized. Some technologies need to be people's source. People need them to know.

What are the complications of such a debacle if it is forgotten? Destruction of a formidable knowledge that, once known, will become unknown. Destruction of ideas that were a vast dream and imagination, only to be erased from memories. Why? What? The most important is when? When are they going to disappear permanently from our lives? This is a question I cannot fathom, nor can it be considered a loss to human civilization. The loss of human recognition. The loss of will to propagate and progress. To go beyond what our minds can do and can bring will be forgotten.

I fear this loss, that such ideas and knowledge will disappear, that they will be forgotten. I fear for I fear. I fear it will be too late. I fear that it will be gone......forever. No worries.

We will bring it all back. The fragmented needs to be defragmented. We need to apply it, do it, and bring it all back—not just the technologies, but also the jobs and careers required to make it happen. We will not be in despair. We will not be tested. We will be unstoppable. So, what are these technologies? Here is the list. The first one that I know of is the Hovercraft technology. The technology is so flexible that it can be used on land and water. For the military, it's perfect for amphibious landings. For the civilians, it is about transportation, of course. Do you need a dock or harbor to accommodate such technology? No. You need to make sure that it is on land that can accommodate it, such as grassland, concrete, or a plain area. Of course, it can run from the sea to the beach and your garage if it can accommodate it. And so is the road if it is wide enough. The good thing is that the technology is there, but further innovation is needed to make it cheaper and more reliable. Can it? Yes. We need to ensure that it will be bigger and better for future use.

- Second is Nanotechnology.
- Bigger Aircraft is still better.
- Civilian Submarines
- Simplification of Concepts and Algorithms, especially in Mathematical Applications (Rebellion), to be used as STEM improvisation
- Rubber Tire Replacements
- Mirror Reflections

- Improvisations of the Telescope (It doesn't need to be Bigger)
- Graphic Design or 3D Improvisation
- Landline Telephones Improvisation
- Cooking Improvisations
- Pavements such as asphalt and other means of better roads
- Grasses (It doesn't need to be taller)
- Art Techniques and Materials Improvisations
- Docks or Harbors
- Airports (Terminals and Runways)
- Spa, Massage, and other Relaxing Techniques
- Parks and Recreation
- Forest Growth and Recovery
- Space Technologies
- Coral Reef Growth or Recovery
- Improvisation for the Recovery of Endangered Species such as Animals and Plants
- Road Traffic Improvisation
- Schools, Colleges, and University Campus Improvisation
- Library Improvisations
- Forum Designations for people to collaborate
- Office and Work Improvisation
- New Farming Techniques
- Businesses such as Stores and Shopping Areas Improvisations
- Child Learning
- Sail Technology (Not only in Sailboats)
- Sound or Speaker Improvisations

There are many more, and they need to be evaluated to determine if they need innovation. All of our technologies need innovation. We just do not prioritize them. We need to invest in those technologies to ensure the improvement of society and the continuation of civilization effectively. Without that, it means technologies that are increasingly innovative, like information technology and artificial intelligence, are improving. However, many remain unchanged for many years or even decades. We need to change that. We need to ensure innovation for the rest.

Chapter 17
Something Behind Deep Space

(The Next and the Final Frontier)

I have been intimidated by it since I was young. I have always been engaged by it since I saw the Space Shuttle mission. It is magic indeed. It is intriguing indeed. And I want to be with it. It is Outer Space, and I want to be in it.

God provided us with a limit to stay on Earth, for outer space is a dangerous place. But we broke that limit. We defy physics and gravity. We made calculations so ingenious that we defy God himself. We become almost immortals. We were divine for we conquered it all, but it is time to break that limit —and we did. We are now in Outer Space.

But to ensure survival for months and years in space, we must do more. The International Space Station, or ISS, performed that mission. We not only survived but we thrived. We are becoming a modern society. A space-age society. ISS has performed many survival missions, and I believe it is time to address these challenges in the long term. For we are ready. We know it now. We must pay attention to the information we gather and use it appropriately. We can finally go to space and stay there for months, even years. We will survive, and we will take advantage of what space can offer. We have technology. We must implement them. We must integrate them.

Today, the United States is the leading space explorer. We have been there for decades, but we are still in our infancy. For we are not engaged by it anymore. We are just thinking about the disasters that happened. That deters me. This discourages me. But we must move on, for we need to keep space exploration open for us ... and the next generation.

Space has been engaging. So many strange things. Dark Matter, Dark Energy, Black Holes, Neutrinos, Neutron Stars, Weird Looking Galaxies, Quasars, Humongous stars, weird exoplanets, fascinating planetary nebulas, Super Massive Black Holes, and many... many... more.

These are the entities that are engaging but extremely distant and difficult to study. That is because the only ones who could be the Astrophysicists, Astronomers, and probably Theoretical Physicists are those who use chalk and a blackboard for hours, and often for days, to prove their calculations. These so-called explorers are the only ones who study space on Earth. No one else can except the people from the ISS and Tiangong Space Station, which are made and used by China.

The Chinese are on their way to conquer space. They have the money, the technology, and the will and determination. We are ready to accept and confront them. SpaceX founder Elon Musk will provide such technologies for the United States. He will make some remarkable Rockets. He is already testing a Rocket so large that Saturn Rockets will be considered obsolete, even though they put men on the moon. It will be formidable. It will be God-like. We must be patient, as testing is already underway to get us to the Moon and Mars. The world will be unified. People all over the world will be filled with confidence and enthusiasm. They will be united with both the United States' banner and the Chinese banner. And probably everyone else's banner.

But we can no longer afford to use rocket technologies to go to space. I can now say it is "obsolete," though it is still useful. There are just too many accidents. Too many vital losses. Too dangerous. We must incorporate modern technology, such as another rocket engine or an advanced drone with powerful engines, to be put into space. It must be used again. It must be used repeatedly. The power engine that must fill that power should be Nuclear Fusion. No other choice can be made. Although we do not have it operational yet, we must keep improvising to make it possible. To make it worthwhile for all humanity and, of course, for space technology itself. I will discuss Nuclear Fusion more in the following chapters.

Remember, we are still in infancy in space itself. There are unique resources that we can incorporate and exploit. We must find a way to obtain resources like Helium-3, which, of course, can be used for Nuclear Fusion itself. Such resources are abundant theoretically on the Moon itself. Just a small portion of it can empower not only the United States for one year but the entire world. We have to go there to mine it.

Another unique resource that may work as well as anti-matter. The world can provide antimatter, but only in tiny quantities. We must go to space to find resources because they are more abundant there. It can also power spaceships itself in an incredibly significant way. We need to develop that technology, but it will be in the next two or three decades. Such technology must be advanced. It must work with antimatter, as other unknown technologies can be implemented using it. We do not know, but Star Trek might be possible.

We must keep dreaming. We must keep an open mind and open vision.

Remember, Space itself has an infinite number of resources we can use. We must go there and mine it. That is the problem. The technology we currently have is inefficient and costly. As I said, we need modern technologies to move forward. I am adamant that we will do it in the next decade because, like us, China is also preparing for it. That is another issue that we must press on.

The exploration of space begins in our solar system with the planets, moons, asteroids, comets, and many other celestial bodies. It will likely take decades to explore everything in the Solar System. Still, we might skip it and focus on interstellar travel to Proxima Centauri first, as it is the nearest star with many planets, including an Earth-like one. It will be a good place to explore. We hope the earth is abundant in life so that one day we can finally perceive we are no longer alone. There are others out there waiting to be explored and discovered.

Thousands of years from now. Once we have the technology for intergalactic travel, we will become a society with undefined boundaries. Undefined because we really do not know what we will become. We are probably a different kind of entity with vast knowledge, god-like ideas, and an unimaginative environment. We will probably live in a so-called Type IV civilization. A civilization that can harness the power of the universe. These civilizations are measured using the Kardashev scale – Type I (harness the power of the planet), Type II (harness the power of the star), Type III (harness the power of the galaxy), Type IV (which I mentioned), and Type V (which is the multiverse itself)[8]. But remember, we are probably not alone, so there is competition. And that competition will deter us from being that civilization. But frankly, it is absurd. We do not want a civilization that will conquer other civilizations. We want a civilization that would work and integrate itself with other civilizations.

Why conquer if you can easily unite the whole universe or just the world?

Do I want to be in a Type IV or Type V Civilization?

No, I want to live a decent life—no need to turn us into Gods. And I do not want to be like Him, but I will praise Him. For He is the Almighty. He is Divine.

God doesn't want to be seen.
But he always wants to provide
We always want to be noticed
But the unnoticed is usually noticed by Him

I hope I am a Cat with Nine Lives.
Because I will always sacrifice myself for you
I will always put my life on the line for your life
But don't make it Nine Lives
One is good enough
Just be sure that one is good enough

Chapter 18
Rise Of Robotics

(The Emergence of Artificial Beings)

The Internet Age has been around since the 1990s. It is getting old. Technology is maturing. The worst part is that it is easily copied. The United States is at the forefront of that technology. However, competition is on the way. China, India, and many other countries want to deter or even replace us as the leading information technology empire.

The United States consistently creates new inventions and products that can impact society, the country, and the world. But in a matter of months, sometimes weeks, the technology can just easily be copied. Let us give an example. ChatGPT is a modern technology that may be a game-changer in the artificial intelligence market. The change may lead to a recovery in the United States' technology market. But for some reason, it was easily copied. It only took a few months for it to be copied. That must not happen again. The problem is that the technology is artificial. Of course, it is software.

An application. A program that can be easily hacked or copied. A young child can copy it, but once copied, codes and algorithms must be changed to avoid being sued for patent infringement. But the structure itself is still the same. They need to tweak it a bit to make it their own.

So, what is my goal in this recent technology called Robotics? We all know it is all Robots and Androids. However, there are other technologies integrated with it. Of course, electronics, microprocessors, nanotechnologies, Artificial Intelligence, and probably Quantum technologies will have to incorporate with Robotics. It must be done because it is an extraordinarily complex technology. Extremely hard to copy. Extremely hard to imitate. Such technology will change the lives of many people. It will provide progress to society. It will galvanize the people. Most importantly, it will finally be a new foundation for the world economy. Such technology will improve the economy. It will add more jobs if it is done appropriately (if not done the right way, then we will lose jobs, and the economy will stall). It will create a new direction and opportunity for people to know what they want to be. For it will give them new confidence. It will provide them with tremendous wealth.

Robotics was once just a fiction during the First Industrial Age, which happened in Europe and the United States. It was always used as fictitious marketing to ensure that society would be overwhelmed and believe it was the next important thing. That was during the 19th century. It was still a different society back then when people began to feel more confident. Still, the technology behind it did not exist, as it required a brain (microchip), nervous system (wiring), complex skeleton (composite materials), and more.

Of course, it needs tremendous engineering to do the job. And with that, it requires promising talents and remarkable ideas. Businesses behind such technology need crucial investments to hire those individuals. Right now, only a few Robotics companies are on the list of Fortune Global 500 companies. That needs to change. Right now, China is leading in the Robotics industry and has the most Robotics companies in the world. That industry needs to get bigger. Such technology will create tremendous wealth and progress in the world's economy and society. Such technology must be made.

Remember, Robotics is a very flexible industry. It is not as flexible as software, AI, and Internet technologies, but it can do the job. It will create a significant sector in the market. Why? That is because it can be used in so many different industries. Of course, the Military comes first. Then, civilians with Corporations and Big Businesses will take advantage of it. SMEs will benefit as well. But the most significant contributors to such technology are the consumers. They are the driving force behind the technology, and the demand will be tremendous, making wealth accumulation unstoppable.

Remember, we need to set a limit on Robotics to prevent it from becoming a job replacement. We also need a stopgap to ensure that the innovation is correct and that the right permanent solution is provided. Of course, a committee hired by the Government must be established to ensure such measures are implemented and enhanced. We must be careful with this technology. It will be a wealth creator, but it could cause damage if not implemented appropriately. One such thing is that it cannot be implemented by a thief, murderer, or anyone who will not follow the law. We can use it as a lover or a partner in life, but we need to ensure that humans have legacies or children to maintain a strong demographic.

Once implemented, the technology will be in our society forever. That means it will be here for thousands or tens of thousands of years. Our current technology, which is software and computers, can be replaced, but not Robotics. It is just a near-perfect technology for our civilization. A technology that will give way to other technologies in the future, such as Quantum technologies, nanotechnologies, and many more. But the technology needs a "limit." It must be contained. I believe that even the Pi Formula is probably unlimited. It is perhaps repeated numbers. And I think those numbers are something else. But I can't assess such things. That's because I knew it would take days, or maybe even months, to determine that in a powerful device, which we can also call a Supercomputer. But I believe it's a repeated number. That's because I believe everything has a limit, and I think only those possessed by no gravity have no limit. But that's in my Sphere Theory Chapter and Gravity Theory, together with the Magnetic Field Chapter.

So, the question is how the brain will work in such a complex machine. How will it provide a decision that relates to the environment that it has to adapt to or will adapt to? It needs to function, not just like a normal robot, but like one that makes such a crucial decision. The decision needs to be not only right but also aligned with future choices. It has to develop a brain so imaginative that only one species can do better – Humans. The only issue with a brain that has Artificial Intelligence of such complexity is that it requires more complexity. Remember, it can only perform Analytical tasks. It does not foster creativity, which is crucial for humans because it is their advantage over the rest of the animal world. That is the reason they started a civilization. That is the reason they become advanced. Can we make the robot have it? We cannot. It is so complex that only a God-like being could probably program in it.

So, I am proposing a Quantum Brain. A brain only specializes in Robotics or Human-like robots. It has to make the right crucial decision, but it has to limit itself to goodness, discipline, and trust. So, the quantum brain is not only complex, but the technology is also so advanced that the robot itself may become a luxury, affordable only to a few. Of course, that is always the beginning. Technology products are always expensive when they are still new. Once the technology blossoms and becomes common in society, it will not only deflate the price but also become a commonality. The issue is that it may take a while to be in that category. Robotics is still a technology that requires significant investment. OK, I am out of my mind. I already said that to other technologies as well. Maybe we need to know that we need to diversify. Investors will be very selective about their investments, given the numerous options in the specific technology they need to engage with. In the end, it will just be another win-win situation for everyone. The wealth will increase, jobs will be more demanding, and poverty will be alleviated. So, in my opinion, let's do it. Why not? We are all technology geeks anyway.

Someone on Twitter asked, "In your opinion, which quality is more valuable: honesty or loyalty?"
My answer is: Honesty. Period. It defines who you are. Loyalty can make you someone you're not.

Chapter 19
Limit

(All Needs a Limit.... And Limit must be a New Universal Law)

Just like in the law of economics, we all have unlimited wants, but we also follow the law that we have limited needs. For such an application, which is Limit, provides a rule to break the abuse of nature, philosophies, mathematics, industries, and all other sectors, for we cannot have unlimited needs or offer solutions that can produce unlimited answers. For everything is limited and must always be limited. We should and should not break the law of nature that anything is unlimited because everything is limited, and it must be kept in mind that everything has a limit.

For example, in astronomy, the universe is enormous. Really huge! But is it limitless? No. It isn't, because even if we cannot fathom the size of the universe, it still has a size that has a limit. However, we see it as limitless because of its humongous size. By the way, I am talking about the observational universe, which is the universe that our eyes and very powerful telescopes can see. The unobservable universe is way bigger, and you cannot fathom its size because it's too big[9]. But you know what I mean. Even in its current state, it still has a limit. That's because there is no such thing as limitless.

So, what am I trying to say? What am I trying to apply? If the laws of nature are not unlimited, then everything else must also not be limitless. Everything must have a limit. The law of economics must have limits. The law of philosophy must have a limit. The law of politics (federal (national) or International) must have a limit. The law of businesses and industries must have a limit. The educational institutions that provide knowledge, applications, and ideas must set boundaries. Even the military and the underworld, which is organized crime, must have a limit. Why? That's because society doesn't want someone or something with unlimited entities, power, ideas, and influence, especially when they might apply or abuse them. The key to that is abuse. Because an unlimited entity can lead someone or something to abuse, it is a provocation that may result in disaster.

In economics, we always check for a bubble in the economy. For a bubble, there is a limit. For a bubble, there is no concept of unlimited law. The bubble will never be limitless, and it will pop. Such a concept in economics is a letdown to economists because they cannot identify the bubble, and as a result, they will not be able to resolve it. This means the only way to fix it is to avoid creating a bubble, which occurs when the market economy becomes more volatile. Is it a bubble? Probably. But it's a warning. Economic bubbles are not limitless and must have limits.

Empires also have bubbles. That means they cannot conquer everything. They may win all their battles in a war, but they will not be allowed to destroy most or all lands, for they will not be able to control and defend them. So, limits must be set in empires and for other countries that tried to conquer the lands. Conquering other nations is obsolete now, anyway. It will not work anymore.

The law of science, especially physics, must provide limits as well. Such knowledge is limitless, but in the end, we will regret it. That's because giving knowledge must be limited, or it will lead to mistakes in educational institutions. For example, Albert Einstein's Special Relativity and General Relativity resulted in such good equations that they changed the laws of physics and science. However, we know that the law must be limited to "Good" use. For such a limit cannot be put on brakes. That means they made it unlimited and created the "Atomic Bomb" ... and, of course, the Hydrogen Bomb. Is it a good thing? Of course not. Such entities are so powerful that they will either disseminate or annihilate societies. As he said, the next World War, which is actually just a local conflict, is merely a fight with stones and sticks.

But whoever said that we have unlimited wants and limited needs is perfectly correct. We are just limited in our needs. This means we must follow the new law that everything has a limit and understand that we cannot conquer it all. We are humans, after all. Even God himself is not limitless. But that's up to you.

I hope I am a prophet.
I hope I can see what you do in the future
So that I will fulfill that future for you
You will always be there in the Present
But I am sure the future as well

Chapter 20
Sphere Theory

Why Is the Sphere the Universal Shape?
Why Is It Out of Control?
In The End, it has a Limit.

I am not a physicist. In fact, I did not study physics thoroughly in High School and College. However, I have been a massive fan of the physics world for two decades now. I watched the Astrophysics section on Science Channel, Discovery Channel, and National Geographic. I read science magazines, especially those about physics, like Scientific American, National Geographic, and New Scientist. I am a devoted fan. Because I am a fan, I want to dedicate this chapter to Physics. It is all about Spheres and why they are the universal shape. I want to share my knowledge and hope you will respect my dedication to the field of physics.

It flatters me that a circular object can be controlled. Yes, as I mentioned, it is out of control, but I will get to that moment later. Balls like the ones we used to play, such as Basketball, Baseball, and many other games, are, of course, spheres. We used that shape because our hands can control it, and, of course, in football, our feet can too. A lot of things are spherical. The planets, moons, stars, elliptical galaxies, and probably galaxy clusters are observed. However, due to gravitational lensing, their shapes are undetermined because some galaxies can be irregularly shaped, while others are invisible or have a regular shape. But are they? And, of course, the newly discovered objects. Black Holes. We finally determined that they, of course, exist, but we also found out they are spherical.

So, what's going on? What makes universal objects spherical? What makes the shape special? Why was it created like that? More importantly, why the Pi formula that determined its shape produced unlimited randomized numbers, which, I believe, provides another reason that there is something unlimited, which contradicts my theories about "LIMIT." But for some reason, it started at number 3. Does that mean Triangular can be something too? Right now, let's focus on Spherical.

Again, what's going on? What makes spheres the object that makes up the universe of all branches of science?

First, as I mentioned, it can be controlled. For example, baseball pitchers, who perfected pitching probably in the 1960s, can control the ball well; they produced the best random pitching ever. It makes them the team's best defenders, but unfortunately for me, it also made baseball boring. But it's ok. It is about the ball. How did they do it? It is just the control of the hands. It does not use any technology to provide or assess it. That means it's all about the formation of how the baseball is thrown to harness such randomized throws. Of course, we do it in basketball, too. We can all control the ball, shooting it quickly and easily now, and sometimes, for some reason, no need to fully concentrate. We have to find the basket and shoot it. It's not like before, when we had to focus on it. So, as for football or soccer in the United States, I have to say it: I apologize. More and more players can bend the ball quickly, just like Beckham..... Messi or Cristiano Ronaldo, whoever is the best one. It's up to you. Then, there is tennis, ping-pong (table tennis), and many more.

Then, let's get to the big ones. The planets, the stars or suns, and the solar system are likely surrounded by the Oort Cloud, which resembles a sphere. We really don't know about the Galaxy Clusters because the Gravitational Lens, as I mentioned again, distorts them. Then, the Elliptical Galaxies. Even the most enormous galaxy in the observable universe, IC-1101, is spherical. But for some reason, the spiral galaxies are smaller than them. Hmm. Are there hidden objects that make it spherical? I am not sure. But just like the Solar System, which we thought was flat before, could that be flat too? Or is it spherical? I am just assessing it. Then, there were the Neutron Stars, Magnetic Stars, and the newly discovered object that required dozens of radio telescopes to capture. It took months or even years to "develop" the picture, ultimately revealing that the Black Hole is spherical[10].

So, let's assess the most significant object that we can observe. Of course, that's the planet Earth. It's spherical, with atmospheres, cores at the center of it, land, or crust, or whatever you call it, probably solid layers. Something is wrong; we still can't assess why the weather, earthquakes, and other natural phenomena are causing it. Why? Is it the motion of the planet? Is the atmosphere? Is it the sun itself? My guess, which is a theory, by the way, is Gravity. We studied Gravity, but not the whole thing. I mean the planet itself. The problem is it's too big. But we have new technologies now that can be studied. Probably. So, we need to assess the situation by exploring the natural anomalies that could harm the planet. For example, earthquakes. It cannot be predicted. We do not know why it happens. But some entities can predict. The animals. They know it is coming. They know there is something that tells them it is coming. They are not Nostradamus nor prophets. But I believe they know something that assesses it. Their surrounding are getting heavy. Is it? The best examples are the birds. They probably can evaluate it. They fly. They need to be light. If they experience a hard time flying, then they have to leave. That's just an assumption, though, or a theory. I am not sure about other animals, but it's probably because they know their habitat will be decimated, and they must leave. Then the event occurs, and they are safe. But that's just an assumption or theory. I hope some experts can do a complete assessment and provide a hypothesis and result. So, my theory is that there is a phenomenon inside the cores that may strengthen gravity or the solid core. Remember, the spheres can do randomized events, and gravity is one of them. And they are originating from the core of the Earth. Or is it? Probably, the Earth's movement is assessing it, too. Is it the inner core's gravity, magnetic fields, or magnetic cores?

Then there are the Black Holes. They are in different sizes, but we know they are all spherical. We have to understand that the most oversized solid objects in the universe are Massive Black Holes. They are so big that some are nearly as big as the whole Solar System. I am probably wrong, but my theories about Limit contradict that it can be unlimited in size. Perhaps because it's circular. Remember, black holes have the most potent, enormous, and deadliest gravity in the whole universe. That doesn't mean the biggest is the most fatal. That's because it doesn't circulate too fast. After all, it is just too big. That means the most lethal are probably the smaller ones or the medium ones. Another assumption, and I am hoping you can assess it. It also means that the movement of spheres causes gravity. Not the inner core. Another assumption again.

Then there are the galaxies. Let's start with the spiral. We all know that Super Massive Black Holes or SMBHs reside at almost every center of the galaxies. We all know that every spiral galaxy has an SMBH in its center. That's because we call it massive. And there is Super Massive and there is Hyper Massive. They are the most significant objects in the universe. And Ultra Massive Black Holes are bigger than SMBH. They are ten billion times the mass of the sun[11]. And that's what we should prove. Why is there an SMBH or UMBH at the center of a galaxy? By the way, UMBH is still an unproven word for Astrophysics. It is still early to say that such black holes exist, and there are very few that exist.

Let's start with the SMBH. We all know it was probably small when it was young. But it keeps getting bigger and bigger as it gets older. Of course, everything has to be eaten at the Event Horizon. Everything else has to wait. And they will wait for a very long time. It's probably millions or billions of years. So, while the black hole is getting bigger because it's eating more, the galaxy, or probably a globular cluster, in its early phase, will also become more immense. Before the black hole swallows an object, it must be dissected first. I believe it will dissect it on a smaller scale, so small we can say it's on a Planck Scale—the smallest measurement of an object. Of course, it will also be spaghettified.

That's how powerful the gravity of the black hole is. It is so powerful that it will almost not follow the law of physics—almost, by the way.[12] That means it's almost there. It is so powerful that it holds objects like stars, nebulae, neutron stars, magnetars, and many more on a humongous scale. So humungous that it created a galaxy. The most enormous formation of the universe. The clouds are probably more massive, but they are just clouds.

Let's work more on the correlation between the galaxies and the SMBH. We already know that if the galaxy is getting bigger, the black hole is also getting bigger. But it's just too many stars. Is the galaxy taking the stars to reside in it? Of course not. There are no stars in the open void. But there are clouds. As I said, they could be bigger than the most enormous galaxy ever made. And they can quickly be taken by any galaxy. That's because they are clouds that gravity can easily grab. Of course. So, what's next? So, once the cloud or nebula is taken, for some reason, it starts to develop stars. More and more stars. It could be thousands in one cloud or probably millions. The cloud is big, so it can easily form stars. So, what's next? The stars, of course, will explode into supernovas, and they will turn into planetary nebulae or clouds. Of course, these clouds or remnants of the planetary nebulae will form more stars and more stars, and the cycle will come back again. That's how stars are formed in a galaxy[13]. By the way, it doesn't form in the open void. Can you assess that?

So, are the spheres causing gravity? We don't know because we haven't been to objects that are not spherical. But they might be magnetic, like the asteroids or comets. Another question is whether the spheres are causing this new dimension, which is gravity. The real question is whether it will open other dimensions, like time, space, or space-time. However, in string theory, there are dozens of dimensions. I hope to know more about it because it's also theoretical. Can the spheres finally break it apart and cause new theories? Let's see. I am not a physicist, just for information, but I have extensive knowledge of Physics. I am just not a professional. I don't do calculations, but I don't assess the theories I provided today. I am just trying to prove that a person like me can probably do something different. And random people on Earth can do something, too. Remember, before scientists came about, there were philosophers. They are probably scientists, but they do more than that. There are great ones like Socrates, Plato, and Aristotle who could do wonders in ancient times. And they started the great Western Civilization. The core of the civilization right now. I am just hoping there are people like them now who can do wonders. Who can make a difference in the world? Who can tell the world that something was discovered? I am hoping they can provide those ideas to the right people. Remember, Adam Smith was not an Economist at all. He was not even a Philosopher. He is just a person who can do wonders. And now, he is the father of the new civilization. Our civilization. The civilization that makes us and governs us and makes this world prosperous. Capitalism, of course. The New Economy that makes us. Thank God he existed.

Remember, my aim is to alleviate or eradicate poverty. I hope these phenomena, such as earthquakes and storms, can be predicted to prevent harm to people worldwide. I hope they can also find and predict spheres, as well as provide an assessment. It is just an assumption anyway, and I apologize if some of you are offended. I want you to know that anyone can do wonders. We have to open their mind. We don't want them to be Hitlers. Just be friendly people who can help the progress of society and the world.

Remember, there are other dimensions as well. We need to check and assess because Gravity probably opens them up. That's because Time can stop, but for some reason, it keeps continuing.

And then there's Dark Matter. We tried and tried to find out if they existed. But it's been a while now, and we could not just find them[14]. There is something wrong. Do they really exist? Or is it gravity, which is also called Modified Gravity or Modified Newtonian Dynamics, or MOND[15]? I know Einstein made a significant change that revolutionized our studies of the universe. But he derived it from Newton. That means we have to go back to the assessment of Einstein's theory, which is gradually used, or we might have to assess Newton's law for such modified gravity assessments. Again, it's another assumption.

The study of the law of physics and other fields, such as Economics, Artificial Intelligence, and many others, is becoming more extreme. We can't do that. We must assess that society needs to understand simplicity and understanding. We have to let them know that everything must be simple so that they can determine what needs to be done.

Remember, everything is a puzzle, and that puzzle needs to be solved—not just in one sector but in other sectors as well. For example, we are always using Interest rate hikes to fight Inflation. Can it be done by other random means? Probably something simpler and predictable. And probably not Economics. We need society to work together to fight it. And other things as well.

We are in the Information Age. We can assess things by resolving them together. We have to act and work like it. Society is the key to solving things together.

Finally, we all know explosions do it spherically. What do you think of the Big Bang, the most enormous explosion in the universe? Does it mean that the universe is spherical? The question is, does that explosion have a limit?

I believe that the universe is spherical. Again, it is caused by the most powerful explosion, and thank God we have never experienced it. It is the Big Bang itself. With the vacuum of space, it will not be stopped. But an explosion has a limit, and that is its distance. Of course, it will be a far distance. And I mean far. So far, we must assess that there is an observable universe, and we will never be able to see the unobservable universe. We may not be able to see it for billions of years. It is just too impossible for the light that provides it not to come to us for billions of years. So, it is a big explosion indeed! And that, my friend, is an understatement. I don't even know if that's the right word.

However, the problem with this theory is that I will fight the protagonist of the Dark Energy Theory. That's because I am a heretic to the theory of Dark Energy, and recent studies suggest it does not exist.[16] The Dark Energy Theory is complex, yet I hope it needs more assessment. Again, after the explosion, everything comes to a halt. So, everything is constant, even from far away. However, something is lurking in the dark. The Great Attractor. The one that contracts everything to its core. So, what is it? What makes it do that? I have one thing in mind. It is forming back. The entity that will reignite the Big Bang. But why is it nearing us? Is it supposed to be far away? Is it supposed to be in an unobservable universe? It could be anywhere. And good lord, it is near us. That means we are near the center of the universe. The Great Attractor is the center of the universe. The Big Bang itself. The one who will do the reincarnation. Remember, the Big Bang is a repeated phonemena[17].

But no worries, the subsequent Big Bang explosion, according to my assessment, will be in another few hundred billion years or probably a trillion years or two. I mean, that is going to be something. But I have a hint. If the explosion usually decimates near the center of the universe or even the middle of it, do you think the far part of it is not? That means no. The far back places of the universe are still surviving from probably multiple Big Bangs. They probably saw the Big Bang many times. Who do I mean? Of course, the alien beings. The one that is perhaps the most advanced being that was made. That's because they have been there for trillions of years. That means they are probably Godlike. They are in the Kardashev Scale, which perhaps controls a Galaxy or the Universe itself.

So, what is our purpose then? We, the humans. What is our true goal? It is to survive. That means we have to go there. The farther part of the universe. That's because we will survive there for trillions of years. But we have to do something meaningful. We have to pledge to the most advanced beings in the universe and say, "Let us live, and we will not provide you harm. We will be your friends, and we will never be your enemy." They know it. They understand. That's because they survived for trillions of years. Then, they do communicate with other advanced civilizations in the universe. They might be drinking beer while their kids are playing.

And then, the best part ever. The white hole. They know it exists. They know it is there. But the problem is that they could not find it. So, what creates the white hole? It doesn't just start from being small. Yes, it doesn't begin with being small. It originates from the black hole itself. Of course, the Ultra Massive Black Hole is probably the most enormous black hole. That's because everything has a limit.

And, of course, the black holes themselves. That's because the UMBHs are the biggest in the universe, and it has to disperse what they eat. That means something must be done. It will explode with another black hole after merging the two galaxies. That's because it will not explode while in its limit. It will still be the same. The only problem is that gravity is so powerful that it will not maintain its size. It will not do anything else until another black hole merges with it. The two black holes will merge and explode, turning into a White Hole. When the white hole is released, it will disperse a lot of energy, causing the explosion of the combined galaxies. And that will turn into an elliptical galaxy. So that means the elliptical galaxy was created due to a blast from the white hole. Once the white hole finishes the explosion, it will again turn into a black hole. So that's the reason we could not find a white hole. They only occur during a blast. Again, this is just a theory. I am aware that I may create fault but it is something that I will sacrifice for it.

And, of course, the spiral galaxy will turn into an elliptical galaxy. By the way, it will be so big that it will miniaturize the spiral galaxies. But it will be a sphere. That's because the white hole will not rotate from left to right, but will disperse the objects. Remember, UMBH is so massive that the two UMBH merging into a vast white hole will take a billion years to expel the objects. It will, of course, but the problem is that I don't know how the white hole moves. The only determination I can provide is that the elliptical galaxy is spherical. So, it must not be turning but just disbursing everything in all directions. I mean all directions. That's because the gravity is reversed. It is not pulling you in but pushing you out. So, the objects go out, and of course, the elliptical galaxy goes out as well. The problem is that others, such as the stars, will be dispersing, and for some reason, the clouds, which are the nebulae, are also being pushed out. But the clouds or nebulae are so soft that they push out of the boundary of the elliptical galaxy after the explosion. That means it's all gone.

However, you cannot push out the stars so easily. They are stars, actually. I mean, they have the strongest gravity in the galaxy and one of the massive objects.

So, what's next? The Elliptical Galaxy is dying. It is in its final journey. It's done. It will never reform or reincarnate anymore. That's why the stars are yellow. They are dying too. The problem is that the elliptical galaxy is spherical, so the stars and the planets that form it are probably spherical too. I will say yes. No. Probably a chaotic one. But it's a different solar system. Totally different from the solar systems of the spiral galaxy. It will not move like a spiral galaxy's solar system, but it will be chaotic. But something tells me they will not intersect, be bumped, or probably be gone. The star in that solar system is perhaps not in the center. It's probably at the side, dispersing out from the core, and the planets are probably following it. Remember, these are moving for millions or billions of light-years. It will take a while to get bigger. And it's getting bigger and bigger. What we have to know is how the Elliptical Galaxy was formed. We all know that some experts assessed that the combined formation of the Andromeda Galaxy and the Milky Way Galaxy will result in an Elliptical Galaxy. We call it the Milky Way Galaxy. The question is how? I checked the formation of the Elliptical Galaxy. Of course, it's in an open space, which is the space itself. The only issue is that it is spherical. This means some formations became spherical. And I made an assessment. It is an explosion bigger than a Supernova or Hypernova. The problem is we never knew how two SMBH or UMBH exploded once they combined. Is it true that once the two SMBH or UMBH explode, it will be so big that it is a galactic explosion? Again, it is just a theory, and it is probably untrue. It's up to the experts to assess and determine if it's true or not.

So, I believe when the two spherical galaxies combine, the two SMBH gather what stars, clouds, or anything else are in that combined formation, and the two SMBH will compact them together. It is so small that it could be smaller than a globular cluster, which is a tiny galaxy. And then it exploded in a galactic explosion. It is so powerful that only the Big Bang can surpass it. But the massive elliptical galaxy will take a while to reach that size. That's because the explosion will not occur quickly. Of course, light is faster than the explosion, and it will take a while for such an explosion to spread into an elliptical galaxy. Again, this is just a theory, and it may not work out.

In the end, we have to assess how that white hole disbursed that gravity out. Does it have a limit? Is there a limit on that elliptical galaxy? I guess there is. It's all about the white hole. It will stop once the white hole is gone. And, of course, it stops disbursing. Because of that, the galaxy will not get bigger anymore. It will just die. Let's see. Is that as big as the IC 1101? No. It is still too early. The universe is still too young. It's probably several times bigger than IC 1101. It is the most enormous galaxy that has been discovered in the universe so far. And, of course, it's elliptical. So, the theory is that the Elliptical Galaxy's center is a white hole in the beginning. But that whole hole will turn from a massive one into a black hole. And again, so is the elliptical galaxy itself. The dying galaxy.

Let's get back to the Great Attractor. The humongous place in the universe that absorbs all objects, including galaxies, clouds, and many other matter in the universe. It is a prominent place, and I am not sure why it has that effect. What is it? Why does it do that? And, of course, you are curious why it is doing it. I made an assessment, of course, that it is probably unknown. But it absorbs the rest, including black holes, magnetars, neutron stars, and everything else so powerful that it can surpass them if I presume. If that is the case, I have to find another one before I formulate such a theory that it is true. And I did not. There is only one Great Attractor. That means only one thing. It is significant, of course, and it absorbs everything else. My only theory is that it is the center of the universe. That's because I believe the universe is in a cycle. It explodes, and it contracts. It is the most enormous explosion in the universe. It is something that has to be found. That's because experts said that there is no center of the universe. It has, too. Why? That's because there was a Big Bang. The origin of the universe itself. It is an explosion, of course. And that is the origin of the blast—the center itself. And we should now know that it is the center of the universe. So, what should be done? We must assess that the Great Attractor is the center of the universe if we can find another one. If not, then that is it.

So, what is it doing? As I mentioned, it is so powerful that it absorbs galaxies and is the most powerful object in the universe. The problem is, as I mentioned, it has a "Limit". And that is the limit it can absorb. The problem is, once it is out of that Limit, will it cause an explosion? It is just a theory, but it may be correct. Again, everything I do here is only theorized, and I don't have the assessment or the privilege to check if that is accurate. However, if that is the center of the universe, then it may be the origin of the next Big Bang—the subsequent most enormous explosion of the universe. Remember, we need to verify the correct calculation for when it will explode and whether my theory is correct. Of course, the enormous explosion of the universe will destroy us. And we need to find out. And of course, I already proposed a theory that the most advanced civilizations in the universe are at the far end, where they not only survived the explosion but also lived for billions or trillions of years, if possible, and became God-like beings.

And, of course, that Great Attractor will become another Big Bang and recreate the universe. Once such an explosion occurs, it will again be another universe in a cycle. The problem is that we have to know the average of the universe. I believe the ages are close. That means it doesn't have a standard age, but it might be close enough for us to develop a theory about when it will explode. However, the explosion will have a limit, of course, and we can also define the size of the universe itself. Can we? It's up to the experts to do that. I will give you a hint. The Elliptical Galaxy is an explosion. And it defines the size. Is the universe that too? Probably. The problem is, we do not know why it does that. We do not know what it consists of. Of course, we have to know. But we don't know about Black Holes, Magnetars, and Neutron Stars first because they are farther away. They might be more valuable than this one. And we may find a way to create a wormhole, making long-distance travel possible. But the Great Attractor is that, too? Are they all part of Gravity? They might be the Gravity itself in a solid form. And, of course, they are so powerful that they will break the space and time continuum. We need to understand how they work and identify the materials or components that are missing from them.

Then, while attending an AI Expo at the Washington DC Convention Center, I met many friendly and accommodating people. They did not hesitate to greet me at all and were very welcoming. One even mentioned the weather issues, and she was very informative. I was disappointed in the AI GPU that her Data Center was using. It was the previous version. She said the latest ones were sold out, and companies were lining up for them. But it's ok. I asked her if she could do it again because I believe her project is more important. It is about the weather and accurately predicting it with inches of ground resolution, likely within milliseconds. Can it be done? Of course. We are still in the infancy of Artificial Intelligence. It can be done. But remember, we have to prioritize other technologies besides AI. We cannot focus on one technology, and everyone else is either dormant or undergoing a slow innovation process. We need to have them, too. Plus, it will create more jobs. More STEM jobs. More companies. More sectors for the economy. More improvement for the society. More innovation for our civilization. More people will open their minds and contribute to the world through their ingenuity. So, do we need to diversify technology into many sectors? Oh, I forgot. Investors will find it hard to identify the right company to invest in. They will make money too. But in the end, everyone will be happy. Everyone will be satisfied. Ultimately, people will believe that the future is real and will look at it. They need to touch it. It will be untouchable, of course, but it will be seen. It has to be seen—the future, of course. I have always thought that the US Capitol needs to be seen. I always see it. It isn't enjoyable, of course. I mean the building itself. It wants to be seen all the time. And I have no choice but to see it. So are other buildings in the world. The Pyramids in Cairo, the Parthenon in Athens, the Eiffel Tower in Paris, Burj Khalifa in Dubai, St. Basil's Cathedral at Red Square in Moscow, and many others. We need something visible in a city. They are icons. They are the symbols of what we are. They symbolize what the town was, is, and will be. I apologize for the intermission. Now, let's get back to the topic.

While attending the AI Expo, someone mentioned the Tunguska explosion in the Siberian Province of Russia. That explosion was so vast and deadly that it was miles and miles of devastation. The only good advantage is that it is in a remote area of the world, with no people at the site of the explosion. It took investigators days to get to the site. Then, I discovered more meteor explosions in that particular part of Russia. It was not just any minor explosion but major explosions as well. I believe a recent one caused such devastation, even though it never had a land impact. It exploded in the air, but the explosion was so vast that houses miles away had their windows broken.

So, I checked the average meteors that enter the Earth's atmosphere and found thousands of them every day. Then, I assessed that if there are thousands of meteors, there are probably a few that are big. They should explode. They don't. They just evaporated in the atmosphere. This is when I found out that the Siberian atmosphere was unique. The meteors that go there do not disintegrate[18]. So I assessed that the atmosphere is soft and the gravity in Siberia is likely similar. Then, I evaluated that the Earth's atmosphere, or any other planet's atmosphere, is neither standard nor irregular. That means we have what we call an irregular atmosphere or irregular gravity. That is just my theoretical assessment. So, even though the whole Earth is spherical, the atmosphere, just like the Earth's crust, is irregular. Probably the Mantle and the Cores coincided with it, too. Earth's crust will be my topic in Seismology, which is the study of Earth's crust, mantle, outer core, and inner core. But I may only talk about the Earth's crust because it is the one we are standing on right now and the one that causes earthquakes and volcanic eruptions.

So, what is the significance of such an irregular atmosphere? That means some atmospheres are so soft that we may not need a rocket to propel a spaceship to space. Can we use a plane that has a jet engine? Probably. But it's not me who can answer that. It is a country that has a unique soft atmosphere. Of course, it is Russia. However, that specific area is remote. There are some roads, but they need to be improved, such as expressways, to accommodate companies, governments, organizations, and more in starting a new business – SPACE EXPLORATION.

There will be space explorers, such as astronauts, cosmonauts, taikonauts, and many more, when one hundred countries or more engage in the space business. There will be space tourism, with companies like SpaceX, Virgin Galactic, and many more. There will be private companies, and SpaceX will be the primary beneficiary and probably one of the major companies. There are probably many more businesses that may begin as start-ups, leading to an explosion of companies. Russia probably needs to regulate them because they will be very aggressive. Of course, Russia will be the primary beneficiary, but the rest of the world must agree on the rules in that area.

But I believe Russia will have a significant say, and they will eventually be the ones who are influenced by the rules, for it is their land.

I only have one favor for that country. I hope they will not favor the "military" in such a business. Why do we need to put the military in space? What needs to be protected or defended? Do Mars' Facilities need paramount security to engage against someone else? That is my proposal, and I hope the country of Russia and the rest of the Unified World will follow my suggestion. It was the protocol during the Cold War when the USSR, now Russia, agreed with the United States that no military engagement would occur in Outer Space. I hope everyone else will follow that protocol. And I hope they will do that as well in that particular part of Siberia that needs to be neutral to ensure that no military engagements will be made against hundreds of different countries around the world, companies, organizations, and many others, and that is where the United Nations will provide that and not Russia. I hope they do that to Jerusalem, too, but that is another topic. That means the United Nations needs to be bigger but more transparent.

So, what is the reason for such a soft atmosphere in Siberia? Remember, in my chapter on Russia, the Siberian Traps were once a vast expanse of lava stretching hundreds of thousands, or possibly millions, of miles due to a massive volcanic explosion. It causes the crust of the Earth to be full of magma inside. It makes it almost flat or plain. No mountains were made after that. By the way, an explosion occurred more than 250 million years ago. There is still magma in the Siberian Traps, but they have not been active since the major explosion 250 million years ago. Does that mean the magma deep inside that crust made the atmosphere in Siberia soft as well? Again, I am not an expert in this field, and I invited experts from around the world to test my theory. It may be correct, but it may not be as well. All I want to tell everyone is that there could be others as well. Can we check Australia, North America, South America, Africa, and many more? The question is, how big is that soft atmosphere? Can we put one there, too? Probably, but I hope it is big. That's because if it's small, then it may be hard for a spaceship to enter that particular soft area of the atmosphere. Spaceship pilots need flexibility and ease when operating spaceships. Of course, it might be small or medium-sized, but in the future, it will be bigger and broader. We need a big, soft atmosphere for that. So, Siberia might be the ideal location if we could not find another place to provide that space for a spaceship or "mothership" in the future. However, it is doubtful. You don't need to build a mothership on Earth if you make it outside Earth in space. It is easy, and you do not need a big engine to propel such a mothership toward outer space if it can stay in outer space.

That means we don't just need Space Facilities in Siberia; we probably need one in the upper atmosphere of Siberia in outer space itself. It is going to be another city in outer space. Seriously, that should have happened for accommodation and other uses, of which I could not comprehend how many there were. There are many different uses, probably in the thousands, such as tourism, space hotels, residential, and many more. There is also space exploration or space laboratories.

As I mentioned again, the theoretical assessments I made for this chapter are highly imaginative. It deserves contradiction because many of its claims will be false or untrue. I did not provide further evidence or studies regarding this matter. However, I want you all to know that I am doing it not just for fun. I want to open people's minds and see if they can make a probable function on how the universe works. Remember, finding another Einstein has been long overdue. Was he already born? We do not know yet, but we have to be optimistic. He will come to us one day, and we have to accept him. He will change the world and make it better. All I want is to open his mind. And also, yours.

There is what we call the Middle Income Trap. I am confused about that. How does it work? I believe Countries have a chance for high income. The Middle Income Trap must diminish.

A Major War is now obsolete.
It has been obsolete because the major powers have made it obsolete
The major powers are not a threat
That's because they are major powers now
The ones we have to worry about are the smaller or unseen ones
If not seen or heard, they will be bigger in a destructive way

Chapter 21
Balance

(Nature can Balance Itself. Can Humans, too?)

Can nature balance itself? Not really. But living specimens can be in ginormous numbers if they have near-unlimited needs —for example, the American Bison. During the 1800s, it had a population of approximately 30 million to 60 million, a state that persisted for centuries[19]. It isn't like that anymore, but some numbers are still left. But was it balanced? No. The habitat provided by the Midwest Plains was so abundant and adequate that it supported such numbers. But it was the Midwest plains back then. Nature provided them with a balanced number that is quite impossible even now. Thanks to the historical facts, it exists. There are other species as well, such as the passenger pigeon. Once the population reaches 3 billion to 5 billion, it is considered extinct[20]. Oh yeah, they darkened the sky, so people thought it was nighttime. But we really don't know because they are now extinct.

But what is this all about? What is my prelude?

Nature can be abundant, flourish into ginormous numbers, and sustain a habitat that can be provided to them. It is because nature provided it to them. Of course, they were decimated by humans, and there is no turning back. But that was back then.

But of course, I have to provide a theory or hypothesis to support my intention. That's because I am not only talking about nature. I am also talking about humans and society. It is currently out of balance. It is not because they are trying to dominate the world, which they have been doing for 10,000 years, but they never sustain it. That means they never keep it in balance. The world needs balance, and humanity must provide it to maintain order and stability. Of course, there is pollution, environmental destruction, habitat loss for animal species, and the decimation of nature itself. Humans are a dominant species. But do they have to dominate to incorporate their wants and needs? Of course not. They need to have a sustainable process to diffuse it.

Balance of Nature is a complex process for undetermined outcomes due to human actions and inactions leading to its demise, loss, and destruction. Such a method is still ongoing. For example, the Amazon Forest has gradually lost its jungle due to human greed. Of course, humans consistently implement their wants but never justify their needs. They always exert such power over nature because they can, and they will. Ultimately, nature has no choice but to fight back against natural destruction. I believe in climate change, but I don't think it's a human consequence or a natural phenomenon. As I mentioned before, during the reign of the dinosaurs, the Arctic and Antarctic were a tropical place with no snow or ice. It was habitable for everyone, so many species from that time, as evidenced by fossils, are found throughout those areas[21]. And there are many. That is my theory of climate change, and I believe that such a course is believable. However, climate change can be determined by human intervention itself. But, for me, it's just too big to formulate such destruction. That's because pollution cannot be a worldwide phenomenon. Probably just minuscule. But that's my hypothesis, and I hope you will respect it. That doesn't mean it doesn't help climate change because it does, but there must be something bigger that incorporates it.

Do we have to worry? Of course, we are. However, such worries must be addressed with a plan and implementation to achieve a good outcome. Is renewable energy good enough? Of course. It is. It is also sustainable. Maintaining renewables is not formidable, and it is eco-friendly.

Ultimately, we need to provide a sustainable, environmentally friendly power source that balances protection for humans and nature itself. Of course, I am not talking about Nuclear Fission, but Nuclear Fusion. The power source of the sun. The power source of all the energy in the universe. The problem is that we cannot implement it sustainably. That's because we cannot determine the primary source of such power or how it is implemented. Which, of course, is gravity. It can be implemented in a gradual process. We are revolutionizing it to manifest gravity or electromagnetism. But it's a complex process. We need to find a way to incorporate gravity or electromagnetism, but we don't have that technology yet. It doesn't mean we need to find a way to do it. We need to find a way to do it without incorporating gravity or electromagnetism. That means we have to find a way to study gravity or electromagnetism and how to implement them. Of course, nuclear fusion will work without gravity or electromagnetism, but it will not work out the way we want it to. That means more power is being incorporated than being the outcome. But there is already a way to do it, and it's not that sustainable yet, but it is a good start. And they will do much work and innovation to provide such work[22].

Gravity is a balance of nature. And that must be worked upon. To ensure more open innovation for humankind and society. But there are more. Because Balance is, of course, a complex process, you just have to provide such a way to continue the balance, or else the balance between man and nature will cease to exist. Remember, storms, floods, and many natural disasters are getting more destructive. I hope it's just a natural phenomenon, but it seems that is not. I hope it will not be because it is still in an early stage, and we don't know what will happen in thirty to forty years. We cannot tell, but we must provide a process to deter it. It is hard, but there are always clues and steps that can help. But there is always one way to stop it. And that's human cooperation and actions to ensure a good outcome.

A Creation... it has its upsides and downsides.
Yet we must put it in our minds that it has a meaning
It has a role after all

Chapter 22
A Must For The Philippines

The Philippines – my country. My home. This was the only place on Earth where I felt true wonder and enlightenment. All I felt for this country is my pride being filled with joy. And I embraced this country because it is not just my home, it is my motherland besides the United States.

Because of that, I want to put something in that will make this country stronger. The country's economy has been growing exponentially since 2010. The average is more than 6% GDP growth. And that is breakneck growth. A growth that may never come again in 10 years. A growth that has to be exploited. The growth that has to be embraced. Because of that, the Philippines needs a humongous reform. It requires a change. It should be leveraged because of its fast growth.

So, what does the Philippines need?

The Philippines needs to work with the investment being given by the investors with the Foreign Direct Investment or FDI. It needs to be leveraged by the substantial amount of money being invested in it. That's because poverty still exists in the country, and it has been long overdue.

So, how can we help the people in poverty in the Philippines?

There are several options available to begin with. The first thing I thought of was starting a local technology company. For decades now, foreign technology companies have been dominating the Philippines. That's because the Filipinos are very smart people and they have STEM professionals who can work for the Foreign Technology Companies. The problem is the country doesn't have one. We cannot wait anymore. Local Technology companies must be build.

To achieve this, the Philippines must prioritize increasing the country's investment and ensure that local banks have the funds to support entrepreneurs or start-ups looking to establish a technology company. Of course, the start-up needs resources to ensure their availability, and because of that, it will have no problem making its products.

Incubators and Accelerators are needed as well. They will guide the start-up company to make it formidable, well-funded, and easy to do business. Of course, the goal is to ensure the success of the start-up.

Another one needed is Venture Capital. The country needed one. We can get one from other countries, but this country needs to have a dozen or two. That's because they will fund the start-up company extremely well, leading to its success.

The Philippines also needed an environment similar to Silicon Valley or Shenzhen, where technology start-ups or firms collaborate to produce outcomes or products that deliver significant profits.

Another one that needs to be done is the relationship between the government and the technology start-ups. The start-up needs Government funds, tax breaks, and many others to ensure they will have ease in their business.

Of course, the start-up needs STEM graduates who will provide productivity to their firms. They will ensure that the start-up succeeds in their business and that the employees do as well.

Schools and Universities Also play a critical role in start-ups. For example, Silicon Valley flourished due to the efforts of Stanford University students who ensured that their peers would start businesses near their turf. That's also the same with Harvard University and Massachusetts Institute of Technology in the Greater Boston Area. Other countries like China and India are following it as well and are producing remarkable results.

The most important factor in a start-up is to turn it into a Unicorn – a billion-dollar firm or start-up. It will make the start-up more recognizable and will lead to more success with its funds. The only issue is that most of the Unicorns do not produce profits.

And then the last one. Turning them into an IPO, or a public company, will drive them to more profits. More capital. And more employees, especially in STEM fields, will work for them. That means the technology start-up succeeds. It is probably a big company or a Corporation. That's how you make the Philippines a formidable country with start-ups.

However, you have to know that there is more.

The Philippines' Land Transportation is a problem in the country. That's because there are no expressways for most of the towns and cities. There are dozens of expressways in Metro Manila, but none in the provinces. They need to build one. Remember, to get rich, you need to build a road.

What's more important about that? Logistics for businesses need better roads. They need their transportation to be fast and easy. They cannot be stuck in a traffic jam. Expressways are necessary to ensure that. It will be good for business to flourish if their deliveries are transported quickly.

The Philippines needed a lot of Subways or Metro Rails. They required a lot of them, especially in the capital, Metro Manila.

Remember, all the biggest cities have the largest Metro Rails or Subways in the world. Cities like Tokyo, New York, London, and Paris have Subways so big that people don't depend on their vehicles anymore.

Remember, we are tackling the Philippines' economy and poverty. What the Philippines needs again is a manufacturing sector. Remember, the Philippines has a huge Service sector but lacks a manufacturing sector. That must change. That must be prioritized. That's because the manufacturing sector can create a lot of jobs. Millions of jobs will probably be created. And in turn, they will become the middle class of the Philippines. It will be a significant change for the country, further driving the economy.

What's next?

Natural Resources, especially the oil resources, are probably in the West Philippine Sea. Some said that we have $16 trillion of oil deposits in the Philippines. We must prioritize that. The Government has to make revenue and profits. Oil is still a big business, and the demand in Asia alone is substantial enough to generate profit.

That's what I believe in the Philippines. We can make a change in this country. That's because it is not only my home but also my motherland. It is my blood and bones. My spirit is within it. So as my mind.

Chapter 23
Fashion Code

(It has been the same since the mid-1990s)

Since the middle of the 1990s, we have been dressing in the same dress code. The same t-shirt, pants, long sleeve, jacket, and many more. We have been using the same fashion code for decades now, and it seems no one noticed but me.

I had it. I wear pants from 2002, and they are not obsolete. It was loose, cocky pants. Did people notice? Nope. They believed it was trendy pants. They even assumed it was new. Did I complain? Nope. It's old and I believe I bought it in 2001 in the Banana Republic store. It was a high-end store, and the pants still looked new even today, after more than 20 years. Probably, because I do not wear it all the time. That could be the reason it is still new.

The problem with that? My pants are still trendy even by today's standards. I can wear it, and nobody cares. People will not complain or make fun of me. Some even think it is good-looking pants.

So, what is the problem with the fashion code?

It has been the same for almost 30 years. We need to change it. It has to be technologically oriented. It must be filled with new designs, fiber, cloth, and everything else but it has to be different. Of course, it has to be washable too. It should be compatible with the washer and dryer.

I am concerned about the fashion code. We have been wearing the same clothes for almost 30 years. Why is it not changing? Is it because Gianni Versace died? Or are the fashion designers getting old? Or we just do not innovate anymore in clothes and technology, it may be useful to. We can do it. We can change it. We just need to be innovative. Right now, Zara, LVMH, Uniqlo, and many more are not operating as they usually do. Their clothes are still the same. They need to innovate. I would love to tell them, but they should know.

So, we need a new fashion code. We need new clothes designs. We need a change in our clothing. Yes! We can. Yes! We can. The Malls need more Apparel Stores anyway. The malls are still very reliable.

Chapter 24
Africa

(The Perfect Integration of a Group of Countries. They Need a Fuel To Do It)

Africa. It is considered one of the most desolate economies in the world because societies are uneducated, bewildered, economically isolated, and, more importantly, unbalanced with chaos and anomalies. The blood of the enslaved people originated on that continent. Undetermined wars occur like bubbles. And more importantly, and that's unfortunately, a wasteland. The Barren Desert is in the North, while the South is fertile, economically ready, and has a young demographic. Sub-Saharan Africa. This is what the chapter is all about. We have already provided such a fictional assessment to the North, for that's the only way to be economically ready for economic integration. But the South is always prepared. That's because it has everything. And more importantly, the geographic shape of the continent makes it more rewarding, for its geography resembles one Genius Psychological Equation – The Gestalt Law[23].

Gestalt Law provides an equation of closeness, similarities, proximities, continuity, closure, and common region. They are the six Gestalt principles or laws. Then, check the continent of Africa. It resembles a near-perfect, probably a perfect one. This means we have new social and economic principles that resemble that law, which may lead to a near-perfect integration of people, culture, economies, and societies, and, most importantly, the unification of the region. Of course, we must build expressways, rails, and everything needed to integrate the economies. Of course, massive economies will be high priorities, but the advanced ones will be as well. That means everyone will have to compete, cooperate, and, of course, work together to achieve what is likely the Perfect Economic Integration that will outperform everyone else. The whole of Africa will be unstoppable. Of course, the countries are already starting with economic development. They need to integrate. That's the one anomaly that they need to work on as a high priority.

Of course, Africa is not a good continent for shipping. We all know that. So, everything has to be prioritized on land. Remember, the Safari is still one of the best natural wonders of humankind. We have to protect that. That's because it's one of the last natural wonders. And I hope it will stay that way for thousands of years. And I am not only talking about Safari. There are others as well. Safari is the most popular one. Big countries like Zaire, Nigeria, South Africa, Angola, and many more must be prioritized, for they are rich in natural resources. Blood Diamond mining is still around, but it is dissipating. But we really don't know that. We have to ensure that women and children will be treated equally. Cultures and traditions are among the things they need to retain. Africa has one of the best in the world. They cannot dissipate, demolish, and destroy their culture.

Remember, the Japanese, who were once among the most advanced societies with the most advanced technologies in the world, still retain their culture and traditions. And because of that, they retained the one thing they are proud of – their people. That's because they discovered who they are and who they will become in the future. The same society. That's what I want Africa to be. That's what I want the world to be. That's because we cannot lose our cultures, norms, traditions, and other aspects of our identity. Too many civilizations and empires lost their histories, but they never lost their cultures and traditions. It is still there. That means we can disintegrate the countries but not vanish what they are for. And that's their traditions and cultures. Too wrong, the Aztecs, Mayans, and the Zulus lost it. I am not sure about the Zulus, but I am hoping you guys can find just a bit of it.

Here is the crucial thing. We know the Continent of Africa is the future. They will be unstoppable. They are thirsty for it. We have to guide them to the right path. We have to ensure they will be in that category – the advanced society, for they are willing to sacrifice everything to do that. We have to provide them with wealth and investments, and most importantly, ensure that they are distributed in the right way – that means corruption must be diminished. The most destructive anomaly is not only in developing countries but also in developed ones. We cannot prevent it, of course. There is no such thing as 100%, but I hope it is 99% if you know what I mean.

God bless Africa. I know you are doing it now. Do more, and you will not just be a behemoth but The Behemoth. Over and out.

Live a lavish life while others live in rot
What have we become
We are not as worthy as a rat

Chapter 25
Integrated Sciences And Technologies

(All Must Flourish But Must Be Contained)

Technologies. They have been there, introduced, and come in different varieties. But we need more. We do not have to introduce them today or in the next decade. But in the coming decades, they have to blossom, integrate, and become one.

(Remember, these are just visions. It is up to you if you want them to be a reality.)

These technologies were once tolerated as pseudoscience. Now, they are a near reality. They will change our future. The technologies I will be introduced to will probably not be made. But I am just offering a visionary outlook to ensure that someone who reads this will be inspired to imagine, curtail, and penetrate the mind, opening it to do wonders. Yes, we can do it. We have to be optimistic that we can do it. We can be different. We can be Gods in a humane way.

So, what are the technologies I am talking about?

First is the integration of Robotics, Nanotechnology, Neurotechnology, and Biotechnology. The power to create superhumans, yet still humans. The power to create super robots that are decent and cooperative. The power to create super soldiers, yet honor and discipline are always on the mind. The power to create a God, but that is too much. I do not want that at all.

Second, Nuclear Fusion technology, more advanced Aerospace technologies, and Advanced Thermodynamics will be integrated to create an Interstellar Spaceship. These technologies will create a ship that will be nearly as fast as the speed of light. They will also have the power to create a new home outside Earth, where a population of more than 10,000 to a million people will live.

The third and final is the integration of All Technologies. It will create a new world or use these technologies to save our planet. Probably, please change it to become a new world. A new Earth where extreme weather, earthquakes, volcanic eruptions, climate change, floods, and many more natural disasters will vanish. It will be the Earth of tomorrow where society will live in peace and harmony. Where Economies will be integrated and flourish more. Where sickness and dying will be non-existent, immortality will be forever. It will be more than a utopia. It will be heaven on Earth. I am hoping it will be in the near future. Just make sure that there is a limit. Society is unpredictable, and we must contain it, but freedom and harmony will always exist.

As I said, this is just a vision, but it will also be a guide. It will be God's choice. But I fear him. I must make the right choice—the choice to seek imagination and implement it to become a reality. We are made for we have these gifts and talents. We must use them to define progress, to define Him, to accept Divinity, to accept the future, and to accept your future, to accept everyone's future.

Billions of years from now, if humanity survives, we will all become God-like. That's our goal. Not to be like God but to serve him. And to serve him, we have to provide him the best acts and thoughts of all – Exemplarity and Creativity.

Chapter 26
Hydrophysics

(We cannot control the Weather, But We cannot turn it into Tier II or Tier III)

I have provided the so-called Sphere theory and what a sphere does. We all know planets are spheres, and their number seems almost unlimited. Of course, everything has a limit, and so do spheres. But there is something about planet Earth, for it is a sphere, and it is just too perfect. The climate, the weather, the air, the wind, and almost everything seem to be bound to every organism's surroundings or habitat. It is so perfect that flowing water changes how the Earth manipulates its movement, surroundings, and, most of all, the flow of other entities like air, wind, and many more.

So, is water the main culprit in Earth's perfect balance of nature? Of course, yes. Water is the main culprit in everything. It controls us. It flourishes us. Of course, it can destroy us. And it can ruin the whole world if nature perceives the world as hostile. Is it Climate Change? Yes. So as Ice Age.

So, what can we do about water?

We all know it is in the ground, underground, in oceans, and in the upper atmosphere. It's everywhere. The problem is that we only harness the resources available on the land. The rivers, lakes, and many freshwater bodies. But can we do something about it? We need to find out how we can harness it. Of course, it's the clouds. It's the rain. It's every cloud in the upper atmosphere.

The good thing is we can finally find out why the storms occur.

It is usually near the equator, but due to the tilted Earth, its storms change every season. It has to because the sun's heat warming the water causes a storm.

But there is a place that experiences rare storms because it's in the Southeast Pacific Ocean. It never has a storm there. The South Atlantic Ocean has a rarity of storms as well[24]. But the lower part of the Northwest Pacific Ocean has a lot of storms, and places like the Philippines and Japan are getting decimated by typhoons and storms all the time. The same with Papua New Guinea, Australia, or the South Pacific Islands near it, such as the Solomon Islands.

Let's get back to the Southeast Pacific Ocean. It has a rarity of storms. That's because there is one culprit—the Peru Current or Humboldt Current. The Peru Current carries cold water from higher altitudes to the equator. That will make the sea surface temperature too cold to support the development of tropical storms[24].

The waters in the South Atlantic Ocean are cold as well, which causes tropical storms to be rare there. And the main culprit for that is the Benguela Current. However, there are other culprits, such as the vertical wind shear and a weak Inter-Tropical Convergence Zone (ITCZ)[24 25]. That will make it difficult for tropical storms to develop appropriately. Just to let you know, ITCZ can create strong winds, torrential rains, and storm surges, which can lead to cyclones.

So, what am I aiming for with these statements that I just

provided? That means we can finally find a way to stop or avoid a storm or even a typhoon. Of course, we cannot stop a storm, and I don't need to tell you about typhoons, too, because they are just too powerful. We all know that. The good thing is we can stop a growing storm or a young storm. Warm air or cold current is the weapon? I am not sure. I am not an expert. I am not a meteorologist. I am just a luminary. A philosopher. I am providing a theory or hypothesis. But what I am trying to say is a logical way to stop a growing storm, or, much worse, a typhoon, a hurricane, and, of course, a cyclone.

Do we need a supercomputer to assess which young storm will become a storm? Probably. That's because there are probably too many young storms and only very few will grow into big storms. Of course, we are going to use a full arsenal of weaponry to stop that young storm: planes, ships, satellites, and everything else. And we cannot do too many young storms. It has to be specific. And we should know that it is the one that will grow into a Super Storm.

Can we do something else too? How about the region where the storms are born? Can we do something about it? We will probably install some entities that will make the storms stop developing. Can we? I am not an expert, of course. But we can probably.

And then, I found another assessment. I found out that cooler water can prevent or dissipate storms altogether. What I am trying to say is that cold seawater in a field as large as the Pacific or Atlantic Ocean is needed to maintain its coldness. It has not been done yet. But I may call it the Grand Mega Project. A Project so large that it will take an ocean to do such. I will tell you what I meant. First, we need to ensure that the sea or the breeding area of the storms is the starting formation, which is nimbus clouds. They are usually the ones that begin a rain. We have to monitor it well. That's because we need to ensure that the region is the site of rain or storm formation. Once we ensure the location of the formation ground, that's when we implement another step. And that ensures that the water is cold in the formation ground. The paths of the storms must also contain cold water. What I am trying to say is we need to ensure that the ocean is cold. That's because once the water is cold, the storm itself will not get stronger. Due to the cold water that leads to a cold climate, the heavy rain will dissipate. Just like the one in the South Atlantic Ocean, it is heading to South America, specifically the Amazon Basin. The area is experiencing the hardest rain all the time. It will dissipate from the rain, and the Amazon Area, or probably almost all of South America, will not experience any more rain. It is just going to be something. But rest assured. It's the Amazon. The jungle needs a lot of water. That means it still needs a lot of rain. But not that much that it causes floods. That means we have to control how the rain is made. That means we have to compute how much rain is going to the Amazon or any landfall on Earth. But I found out that a cold climate, created by cold water, can cause heavy rain to dissipate. That means we can stop the heavy rain. That means we have a fix for the problem affecting Amazon and most of South America. We need some planning and implementation. I will get to that later in this chapter.

That can be fixed in India as well. The storms or heavy rain start in the Indian Ocean, the Bay of Bengal, and the Arabian Sea. This means the region of the storm's formation is smaller compared to the Atlantic Ocean and the Pacific Ocean.

Ok, now that we have a fix for the Summer Weather, it's time for Winter Weather. OK, to tell you the truth. It's all the same. Of course, we know the fix. If Summer needs a cold climate for equilibrium, it helps equalize or normalize the weather. Then in Winter, we need a warm environment to normalize the winter storm. Guess where the Winter storm originates? In the Seas or Oceans of course. It is just too destructive inland. But is it dangerous to the seas or oceans? Probably? But it is where it originates. That means we have to find a way to fix it before it escalates into a Winter storm. It can be a 1-inch, 2-inch, or 3-inch storm, but we cannot make the snowfall any higher than that. We need to ensure that the snow will cover the grass but not the pavement. I experienced that a lot, and I have no problem with it. We want to see snow anyway in Winter. I have lived in Washington, DC, for a very long time. We are safe here from the big storms. That's because the mountains in West Virginia block it. Just like the Andes mountains blocking the rain in South America. But we always have the most significant storms, and that one needs to be under control or prioritized. I experienced 12 to 18 inches of snow almost 20 years ago. I was an essential employee at that time. That means I have to go to work regardless of the weather. But it's ok to make an excuse. You will get it, of course. But I want to be a reliable employee, so I have to go there. It was a havoc. It's a bumpy ride and a slippery situation all the time. But I saw a Plow Truck heading to my work. So, I decided to be at the back of that Plow Truck. It works well. I thought it was only me.

When I looked at the back of my car, I saw four or more vehicles following me. What surprised me was that everyone on my shift was working. Even the software programmers and some supervisors are involved. They have to be there to ensure that their applications will work in that storm. And that was around 6 AM to 8 AM on Sunday, and the snow was pouring. They have to be there for weekly maintenance (Sunday Window) anyway, and they have to stay because they have no choice. The colossal building has many activities to do anyway. It has a gym, where you can walk all day, and there are plenty of snacks in the vending machine, along with people to talk to. You will not get bored. Plus, there is a big window where you can see a big field covered with a dozen inches of snow. It's a good view. In the end, we were tired, but we were not bored.

So, that is the experience of the winter storm. I am telling you because some of you haven't experienced that before, especially you guys in the tropical region. Winter storms are bad, and they have to dissipate or normalize.

So, who will stop it? Most or all countries in the world are affected. Of course. Nobody else. I guess it will take an organization to do that. Does the United Nations have one? Not yet. But the United States can start and lead a group. Hurricanes are awful to that country anyway. Probably, my country, the Philippines. Annually, it is attacked by a dozen or more storms or typhoons. They are used to it anyway. However, casualties are made, and they need something to stop it.

It is my country too. I will do everything to stop it.

Thanks for the support.

So, how to implement such controls to moderate or ease the weather? Again, we cannot control. However, we can change the temperature of the water to ensure that it will maintain the moderation of the weather. For that, it will require many tasks and a construction scale larger than any mega-project ever undertaken. Giga-project? No. More like a Colossal Project. That's because it requires most, if not all, countries to work together to accomplish such projects. So, I am not an expert in the planning and implementation of such a project. My first proposal is installations. A massive number of them. It doesn't need to be in most parts of the oceans, but it is sufficient to adjust the temperature. The second proposal involves pipelines that can adapt within the storm's starting ground. The pipelines may be circular or designed to facilitate temperature adjustments. Third proposal, and this is colossal. This will be an enormous project. It requires pipelines that start in a country at one ocean and end in another country at the opposite ocean. Remember, I mentioned the country. I mentioned it to the other end. That means we have no choice. We have to work on this together. We will have to ask our other neighboring country," Hey! Can you help us control the weather around the world?" They may say yes. Probably.

So, is this proposal going to work? Of course not. I am assessing which projects need to be completed to accomplish this colossal task. Humongous is an understatement. I could not even fathom such. It will be "Colossal."

Remember, we have to work on such a project. It is the only way. Indifference must be abolished, contained, and vanished. It is the only way. We are an integrated world anyway. We are One.

I am a light that will travel through eternity,
but you will be the gravity that warps it.
Yet, you will not stop me — no one can,
for I am unstoppable.

Chapter 27
Cuba Must Keep Its Past And Make It A Present

(It has been like that for 60 years. Why not?)

Cuba has been locked in for 60 years since the communist revolution and the embargo in the 1960s. The past has been reminiscent of depression, destruction, the same design, the same concept, and the same style since it started. The cars or the vehicles are still the same. The houses' exteriors and interiors are all the same.

However, people are changing, and so is Cuba's opening to the whole world. Cuba has transformed and has been trying to tell the world that it has changed since Miguel Diaz-Canel became the First Secretary of the Communist Party and the President of the country, replacing Raúl Castro in 2018. Raúl Castro was the brother of Fidel Castro, the leader of Cuba for decades. Raul has been the leader for only 10 years after Fidel died.

Cuba is full of vintage objects. However, the one that stands out is their vintage vehicles. They are just too beautiful, and they have been around for more than 60 years. Due to the embargo, they cannot be replaced with a new one. Because of that, they have had the identical vehicles for decades since the 60s. One thing, though, is that the Cubans can sell their well-preserved cars for a reasonable price to foreigners, but at a cheaper price if they are locals. In the year 2014, Cuba's vintage cars in good condition were worth from $15,000 to $70,000 US dollars[26]. It's still expensive, but foreigners will buy it for its style, design, and, of course, most importantly, its history.

What else can be preserved? Their homes of course. They cannot sell it to foreigners, but they can rent it. People will rent it, and they want to live there even temporarily. However, if you're going to sell that home, you have to marry a Cuban citizen. For the year 2022, A well-preserved home can be sold for $36,000 US Dollars[27].

Cuban people can also sell their home goods, such as art, like paintings, sculptures, and other items. They can sell whatever they want, and they will make money off of it. That is what the Cuban Government wants. They want their citizens to make money.

Cuba must retain its past. It is the only way for the country to thrive again. They will try to open, but the US Government must allow them to open up. I believe the Cuban people have changed. Once they see that the world is improving, they want to be part of it too. They want to be like the Chinese, the Indians, and the Africans, whose lives are getting better. We have to give them a chance. They are not the enemy anymore. They are friends. They are always friends.

Cuba is a one-of-a-kind country. It has all the vintages, and I believe keeping it will increase their profits. Please do not accept new vehicles, objects, and many more. It will destroy the norms, culture, and traditions of Cuba. They can install new appliances in their home, such as a TV, stereo, refrigerator, cooking range, and oven, but they must maintain the exterior and interior design. Of course, the architecture must be well-kept. It cannot be replaced with new ones.

PLEASE, CUBA.
YOU MUST PRESERVE YOUR PAST AND
KEEP IT AS THE PRESENT.
IT IS THE ONLY WAY.
IT IS THE ONLY CHOICE.

I was a wanderer, I was a nomad
You prove me no wrong, for I am now enlightened
I am now loved, for my soul is yours

Chapter 28
We Need An Advanced Seismology

(Need to prevent Earthquakes and Volcanic Eruptions from happening)

I have been thinking about it. I was thinking about what would happen if we finally eliminated earthquakes and volcanic eruptions. Not just ordinary earthquakes but super earthquakes. Not just ordinary volcanic eruptions, but super volcanic eruptions. What if there is a possibility? What if it can be done?

And I believe it can be done.

I believe the Earth's crust and mantle are irregular tectonic plates that will disrupt countless lives if they continue to move. It will not only be destructive, but it will also change lives for the worse. Earthquakes and volcanic eruptions are the culprits. Both will also cause tsunamis or big tidal waves that will devastate lives. It has to be stopped. It has to be prevented.

I found a way, but again, like many others, it is theoretical or hypothetical, but it may work. It may prevent earthquakes and volcanic eruptions in the future. We just have to provide the necessary functions to do the job. What are those? Let me explain.

The first requirement is to build an X-ray beam or something similar. On its own, it doesn't do anything, but if it intersects with four, five, or more of the same X-ray beams, it will cause the relay point to decompose whatever it touches[28]. The device has to be as big as a huge building. It will be enough to make a big ray. We have to make hundreds or thousands of them in different parts of the Earth. It must be all over the Earth. Probably every country in the world, or eight or more of them, if the government is as big as the United States, China, or Russia.

The goal is to intersect them at a relay point. And then, decompose whatever irregularities that may cause earthquakes or volcanic eruptions.

I do not know how Seismology works. I am just a man of theories. However, if they are the experts, they would see the difference. Of course, the X-rays, microwaves, or whatever technology it is can see the irregularities too. However, the seismologists know what to do. They will fix the irregularities and make it normal. And that will stop or prevent earthquakes, volcanic eruptions, or any seismic shift that could devastate hundreds, thousands, or even millions of lives.

Chapter 29
Make Surgery Almost Obsolete

This is another theory that may work. It was proposed by someone who told me about it when I was doing the previous chapter, "We Need Advanced Seismology." It is also the intersection of a laser or X-ray beam that causes the disintegration of the objects. If used properly, it can work well within our bodies. That means there is no need for surgery anymore.

So, what do I mean by this?

If multiple laser beams or X-ray beams intersect, I believe they can disintegrate the object as I mentioned. It will be a good idea to remove small or foreign objects from our bodies. What are those foreign objects? Blood clots, gallstones, kidney stones, bladder stones, cancer tumors, and many more. It is probably beneficial for the pancreas in the treatment of diabetes. It is perhaps a good one to heal the stomach as well. Probably the heart and the intestines. A lot of use.

So, can this work well? I am not sure, but it can be done. We need to be sure it will work well. We need to experiment with it, implement it, and see what happens.

If that's the case, many people will be saved. The blood clot for the blood vessels alone will work well. So, as a Brain tumor. I am not sure about cancer, but will it work against the tumors as well?

We must keep in mind that it is a great discovery. Imagine if this works out. Imagine the use of it. And most of all, many lives will be saved.

I heard it was already implemented, but I checked with an AI and found nothing. However, let us understand its real potential. I hope someone in the science or medicine field will read this chapter and provide an experiment and its implementation.

The only problem is it's inside the body. Of course, we need a powerful tool that can see inside the body. That's what we need to implement as well. A real-time check of what's inside the body, especially the organs. It will work well. I can say a real-time CT scan will work, but it's currently in the experimental stage. With both pieces of equipment, we can make surgery almost obsolete.

Chapter 30
Shenzhen

(The Fastest Growing City in the History of Mankind. It is still a very young city.)

Is it proven? Is it a reality? Is it worth living for? Is it worth dying for? Shenzhen. The city that only a few know. I could ask the whole street in my place; no one would know about the city. That's because the city has a short history. Thirty years of history. The town was once a fishing village of only thirty thousand inhabitants. I cannot just fathom such an occurrence. I have no choice but to put it in my book, for this city bears a remarkable story. It is a noteworthy phenomenon that must be told. That's because the once-fishing village was handpicked as the model of all of China.

Many were handpicked, but Shenzhen is the most special. Its geography is so perfect for formidable economic growth. It is just beside Hong Kong, which, in the 1980s, was part of the United Kingdom. Hong Kong was once the gateway to China when the country was still not open to the world. With Shenzhen nearby, multinational companies were attracted to establish manufacturing operations. With the firms, the demand for jobs was rampant. With that, Shenzhen has become the most attractive place for migrants—millions of them. No one can stop them from moving there because more and more multinational firms are moving there, too.

It has become the most significant movement of migration in the history of humanity. From thirty thousand villagers in the 1980s[29] to twenty million inhabitants in 2022. Of course, it's mostly migrants—migrants who do not know what their future will be like. But many of them succeeded and became middle-class. Their salaries and wages are multiple times higher than the rest of China. Though the standard of living is higher, too, many people can still afford it.

So, what makes Shenzhen unique? Why am I talking about it? That's because Shenzhen is the model for other countries regarding what their cities will be like if they have the proper planning. That's because Shenzhen is just the beginning. Many more cities will follow the same astounding path not only in China but also in other developing countries, such as India. It's probably a town in one of the countries in the Middle East. It's probably in one of the cities somewhere in Africa. It's perhaps a new city in Europe that might be built from scratch—or even in the United States. We have to follow what China did. What made them the fastest-growing economy in the history of humankind? Can it be beaten? We already knew it was done once. Of course, many will follow. We must believe that such a phenomenon can occur, for I think that the rest of the world will rise. They need a model that can give them the right path, directions, and guidance.

However, China is different. It is a country of socialism and capitalism with a Chinese character. The government works as one, for they think as one. They believe that everything will be achieved as one. They believe it can be done. Now, they have the most extensive High-Speed Rail system in the world. It is so big that even the entire high-speed railway system in the world combined cannot match it. Recently, the country has also had the most extensive Expressway system in the world. Another achievement that will never be broken. The government keeps building. And they make it fast. Real fast. They can build a bridge in less than a month. Or a building in a week.

But someone needs to hit the brakes. If they keep doing it, the economic growth will not be feasible. For some sectors, the brakes will hit the brakes, just like Real Estate. It is now one of the declining sectors of the economy. China needs to find a way to deal with it. Demographics play a terrible role as well. As the population continues to work, they forget that they need another generation to continue the path. They need to increase their population significantly. Even the city of Shenzhen is having issues with the growing living standards. For it is becoming unaffordable anymore.

So why am I talking about the country and the city? Remember, they are the model of development. We can follow their paths, but we must not accept the failures they made. The country and city will still grow, but it is more intense now.

Shenzhen will always be in the minds of the Chinese people, for it is their trophy of accomplishment. But we need to ensure that another Shenzhen will be built. We have to wait for it. Once it is built, investors, people, and even enthusiasts will flock to the city, making Shenzhen the center of the world. I even said it when I went there. Again, as I mentioned, it is currently the center of the world. And the next center of the world will be more than Shenzhen. Of course, it is not in China. But no worries, China. You are still the center of the world. Your country, of course. Just don't keep the theatrics. You and your citizens are advanced, of course. But they advanced too fast. And they did not hit the brakes. This means they turned to the dark side. They became greedy. But it's okay. They found the mistakes in the early years of the country's new era of dynamism. That means mistakes, even those so big, can still be fixed easily. That's because it's too big. After all, it happened when society was still in its infancy. Your society is still in its infancy, and they have found their mistakes. They have to encounter their mistakes. It is not because they have to experience the errors. Of course, it's about learning from it. Too obvious, of course. But that's how great societies were formed. Remember, the Great Depression made the best American Generation ever. And they experienced the most significant war ever. And they made the best country in the 20th century. Century.

I am not saying that we must experience recession, depression, or economic turmoil to experience our mistakes. We have to find the errors before they occur. But that's hard. But it will work. But yeah. The reality is. We have to experience it. Yeah, we did in this generation, with the Great Recession. It made us the new Greatest Society in the 21st century. Century. Not only Americans, of course. Not the Chinese, Russians, Europeans, or anyone else. But it is everyone. And with everyone, we will work together as Armada to create an incredible reality that has never existed before. That is the World's Golden Age—the Global Renaissance. I want to experience it. We all want to experience it. We are all Enthusiasts, after all. Of course, we are. We've been curious since we were children. Don't worry, animals too. Thank God they are not humans. Or else....

China will be a great country. We have to accept it. Not because they will challenge us into a competition, but because they will give us repercussions that will benefit humanity. And they are doing it. The problem is, it was just too fast. But the mistake is it was not too fast. They just did not hit the brakes. And that is, they did not see the anomalies. They need to see it. Now, the whole world sees it, especially the emerging ones. They will now find a way to find those anomalies, solve them, diminish them, and, of course, not allow them to come back. Again, China will recover. They have to make sure that it will never happen again. Don't worry, the United States, competing against you with unfair competition is an anomaly, too. They will find a way to solve it. But you can, too. Just do it the excellent way.

In 10 years or more, more countries will emerge. India will be the leader, of course. But the United States and China have to provide them assistance, of course. And so does the rest of the world. After India, another country will probably emerge, somewhere in Africa or the Middle East. We have to help them.

You are now the center of the world. But provide others with that position, too. That's because you are not only the center of the world. There are other cities too. New York, Paris, Tokyo, Berlin, Sao Paulo, Los Angeles, and many more. They are the center of the world, of course. But they need to fulfill the position. Because, for some reason, they aren't. That means they need a light to give them direction. As a Luminary, I will try. But it's the society that has to do it. That's because society is those cities' key, fuel, and battery charger. Come on, guys, it's not the end of the world yet. You still have it. Check the world, see what it has, and open your mind. It will open your mind. No worries. The world is breathtaking. The world is formidable.

India can probably break that record and so as the African continent.

You Can have an Empire of Imagination
You can Have Knowledge as vast as the Universe itself
But in the End
It is about Who You are
Not What You Are

Chapter 31
Myanmar

(The Country That Intersects the Whole Asia)

The country of Myanmar. A medium-sized country with a chaotic history – past, present, and hope, not the future. It has one of the best geographical locations in Asia, probably in the whole world. In the West is South Asia with India, Bangladesh, India, Nepal, Bhutan, and Pakistan, and the rest is the entire Middle East. To the East is the whole of Southeast Asia – the whole region. Then, the North, which is China. Not just China but the essential cities that are important such as Kunming, Chengdu, Chongqing, and most importantly of all, the Guangdong province, which compromises the cities of Guangzhou, Shenzhen, Macau, and Hong Kong, and the rest of the Pearl River Delta cities.

This whole region is divided or not connected because of one place. One country. One Intersection that needs to connect them. And that is Myanmar. The country badly needed to open for it is just geographically perfect. The Junta may not allow it, but it is a must and must be done. The people are ready, but it seems something stops. Was the Junta afraid? Why did they sack Ang San Suu Kyi, who is the de facto leader of the nation[30]? Is she a weakness or a reformer? Is the Junta afraid of her? Is it because she is a woman? Or even if she is a woman, is her power just formidable?

The Junta has been ruling the country for many decades now with strict rules and a closed society. They are not willing to open, reform, or progress. What made them think that? This is probably because of insecurity or to ensure power and influence.

That must change. The country has an astounding wealth of natural resources, especially gemstones[31]. Untapped Petroleum and Natural Gas are there, too. And more than that, the sea is excellent for shipping and many other things. The problem is that only Yangon (Rangoon) is the only city there. Everything else is desolated. But if they open, there might be new towns.

Then, the inland—the most critical region of Myanmar—is where the transportation of many products is active using roads (expressways) and rail (cargo trains). Remember, they cannot transport products from India, Nepal, and Bhutan to the North. The Himalayas are just blocking it...... badly. And of course, China, too. There is no other choice. It has to cross Myanmar.

So, what will happen? Geez.... More cities. Big cities. Cities that will work with the transportation of goods, tourists, of course, and investors. Myanmar has it. They need to open. Are they ready? The new generation needs to study, absorb knowledge, and use what they learn to create a formidable society. Leaders must be made as well. And most importantly, the Junta needs to change.

Is it over for her? No. But it seems she is getting old, but she still can. She needs hope and supporters who will fight for her. She is a woman, after all, but she can be trusted. She needs cabinet members who can support her. She needs all the support she can get.

In reality, the country needs to open up and see what will

happen. Can the government do something to make such a difference? Of course. They need to realize they are the most critical geographical intersection in Asia. That means they are the proper Center of Asia. It will probably be the Center of the World in the future. They will. Just do it.

Tomorrow might be unpredictable.
But with you, I can predict it
For you have the clues to it
You have the clues to see the future
You have the clues to my heart

Chapter 32
North Korea

(A Nation that must unite an already United or Divided World)

North Korea. Check it out. It is in the center of the countries in East Asia. It is the center of the once dark age of war and destruction that separated the two countries, invaded by the UN Allies, and then the Chinese came and swept. It was a hellhole that went back and forth, and the extreme cold made the war unimaginable. The problem is not the war. It is the peace. It has been more than 70 years, and all it has had is a ceasefire. And it only means one thing. It is still engaged in a war. One of the longest modern wars. But it cheated on that. It cheated for the first five years and was bloody; the remaining 65 years were peaceful. So, what is my aim in this country? What is the real reason for this chapter?

North Korea is another intersection that needs to be addressed, another country that needs to open, and, of course, it requires certain conditions to accomplish this. What accomplishments does it need? North Korea is at the center of East Asia. China and Russia are now open to the world in the North. It has South Korea and Japan, which are also open in the South. The issue is that these countries border North Korea, which is a significant problem. Remember, each of these countries has trillions of dollars in GDP.

What is needed to accomplish a closed country and open it to the world? First, it requires good infrastructure, transportation, and very secure communication. Can we say Digital Communications? Yes, they will work, but they have a vulnerability. Once such technology is compromised, the data itself might be taken away. We do not want Analog Technologies, of course. We want digital communication technology that is encrypted but can be decrypted for authentication purposes. That's the key. That's because digital communication is accessible without security, but it becomes way too difficult if decrypted. Usually, it would be best if you had a supercomputer to do such decryption. You hope it will take minutes or hours, but sometimes, it may take days. What you want is its usefulness, which is the delay of the decryption. Can a Quantum Computer function such algorithms to decipher the encryption? Remember, it is combining the binaries of 0 and 1. It is flexible. It is probably too complex. It can probably decrypt anything. Again, what we need is the delay of the decryption. That is the problem of Digital Technology. It is easy, but it is hard if you provide the right technology to secure it. The problem is it always opens a hole. However, the goal of cybersecurity is to close that hole once it opens.

You also need the country's economic upheaval. You want them to open their world. Of course, it will start with the economy. It will start with the trust of the trading environment—the countries, companies, organizations, investors, and many more. It needs to compete but not dominate. It needs to ensure clarity or transparency with the assurance that the provided entity above will not commit fraud in the form of stocks, bonds, debt, and many more financial obligations, regulations, and deregulations that can be accomplished by the US Government.

One of the biggest problems with economics right now that is not solvable is the Oligopolies. Most of the Global Fortune 500 Companies are oligopolies. So as the Forbes 2000 Biggest Companies. Competition is too harsh. There is no opportunity for the new competitor to compete. And that is devastating. That's because it will take another company from another country to compete against the dominating or monopolistic company. No chance for them to survive and of course, thrive. North Korea must deal with it while they are under the reform process. They do not want a powerful private company that has a powerful influence on the rest of the world.

Once North Korea opens, cities near it will flourish. One of them is Russia's Vladivostok. I believe it will be the biggest city in Russia once North Korea opens. The opportunity is just too much. Another one is Seoul of course. Shenyang, Dalian, and Changchun are another beneficiary. Of course, Japan will be the cities near South Korea including Fukuoka, Nagasaki, and Hiroshima.

Japan needs to build a tunnel probably from Fukuoka to South Korea, which is probably Busan – another beneficiary of an opened North Korea even though it is in the south of South Korea. That's because it is going to benefit well from Japan. It probably going to need two tunnels – one in Eastern Southern Korea which is in Busan and the other one is in the West of South Korea. The two countries will benefit so as Russia and China as well. Probably Taiwan as well even though it is far away. That is because the end conflict of between the two Koreas will lead to peace and prosperity in the region. East Asia will again become formidable with China and Japan as the leaders.

Chapter 33
The Landbridge And A Wider Canal

(For The Canal is a 20th Century technology. We need a 21st Century Technology Now)

What will work? A Canal or a Landbridge? Let's assess it. A canal is only one pathway for one ship at a time. In the Panama Canal alone, it will take an average of 11 to 15 days of waiting for ships to just cross the canal[32]. But it worked after all. The Landbridge is transporting the container goods from one end of the ocean or sea to the other ocean or sea. It will be using probably Cargo Trains or Cargo Trucks. Probably for flexibility. That means the country must build probably the two biggest Shipping Ports in the world just to accommodate the Ships in the South China Sea which is dominated by China, East Asia, and Southeast Asia. On the other side is the Indian Ocean which is dominated by India and of course, the rest of the world like the Middle East, Europe, Africa, South Asia, and probably Eastern North America and Eastern South America, and the Caribbeans.

So, let's assess it. Why build a Landbridge in Thailand[33]? Of course, it will make travel faster, more secure, and easier. I mean we will just have to bypass the Strait of Malacca, its pirates who are worse than the Somalian Pirates in the Red Sea. The Strait of Malacca is one of the most dangerous waterways in the world because of these tugs. But no worries, building the Landbridge will bypass them and the Strait of Malacca and the busiest one, Singapore Port.

Yes, the Singapore Port. Not the most dangerous place in the world. The country/city is the number one best place to live. Its society is the most advanced in the world. The technology companies. The most uncorrupt government. The best drinkable water and some come from water waste. And of course, the container port. It is also the most important container port in the world because it is the busiest in the world with all the automation and one hundred thousand workers. A lot of workers right? An important container port. And of course, it will be replaced by the Thai Landbridge. Awesome! Now, international bureaucracy comes to mind.

However, it's only 7% of Singapore's GDP[34]. It is big. But Singapore's manufacturing prowess which needs that container port is miniscule now and some say it is non-existent. So why is it the busiest in the world? That's because it transports shipping containers from one cargo ship to another cargo ship. It has too. Why? Because Countries like China need those ships a lot. Remember, the majority of the busiest container ports in the world are in China and they need those ships back[35]. So, they put those ships in Singapore Port and have others else transport them to other countries around the world. FYI, Singapore, and China are not that near but not far either. It's just the geography is so good that Singapore is the right destination for those ships. But.... It must be replaced.... I mean it cannot be Singapore anymore. Remember, Singapore is a white-collar country[36]. The blue-collar may still work but we probably make them white-collar for society is the most advanced in the world with its white-collar jobs and the blue-collar workers will always be welcome to retrain to become white-collar workers. There are just more than one hundred thousand of them. the good thing is one hundred thousand jobs or probably a million job openings will be in Singapore in the future and those workers are needed badly.

Of course, the Expressway and the Cargo Rail that transport goods from one shipping port to the other shipping port must be wide, accommodating, flexible, and fast. The containers cannot wait even in seconds. For this will be the busiest expressways, railways, and probably the busiest automation in the world. Robotics at the highest level will be there so as Artificial Intelligence. Will there be drivers in those cargo trucks and cargo trains? Absolutely not. Of course, we all know the Shopping Port due to the automation provided by current shipping docks today such as in Singapore and some Shipping Port in China. The automation was so successful they were able to assess almost the top ten busiest shipping ports in the world. Singapore Shipping Port is actually number one and the first one to use automation. Yes, no driver operating the cargo trucks. It's all automation but the container cranes are still humanely driven. They are experimenting with automatic cranes currently. So, once the Landbridge is opened, it will be fully automated with trucks, cranes, and of course, their movements.

However, the Landbridge will be different. It will contain expressways so it will be more complex but not complicated. You just have to build the most complex automation on Earth. Remember, we are transporting ships from East Asia and Southeast Asia to the "REST" of the world via this Landbridge. Transporting goods from India to China and vice versa using ships is already busy and it may be busier in years or decades to come. Remember, we haven't built a Landbridge before. If Thailand's Landbridge is successful, it will be a significant feat. A feat that no one has ever dared to undertake since the Panama Canal. The innovation that will change the way technology works. We have to envision how it will be built, and how it will be for such a place will probably be the busiest in the world.

Southeast Asia is one of the most important contributors to world economic growth. It will be the most critical contributor to the World Economy in the coming years or decades. We have to get ready for it. We have to find anomalies that will contribute to the slowdown of its growth, as well as its integration and unification. There are still countries in Southeast Asia that need help, like Cambodia, Laos, and of course, Myanmar, the previous chapter. We have to find a way to resolve conflicts, chaos, burdens, corruption, and, of course, many more. But to resolve it, we need to make people focus on the vision. A vision that we will make it. A vision that will make us a people – United and focused, and ultimately, an Advanced Society. Singapore has already demonstrated its commitment to this region thanks to Lee Kuan Yew and his cabinets. It took him more than 40 to 50 years to create the most advanced society in the world, but it's actually fast in human historical terms[37]. I mean, just one generation. We have to prove that the rest of Southeast Asia will be in that feat. Myanmar is one key. Thai Landbridge will be the other. There will be more. I can't just fathom it.

Yes. The most important ones are in Central America. This is one of the most impoverished regions in the world. Probably. But they need to make money. They need something that makes them worthwhile. Well, Panama did it with their Canal. Now, they have to formulate the Landbridge for all countries in Central America. Honduras, Costa Rica, and Guatemala are some good options. Others as well, but they don't intersect their land in either the Pacific Ocean or Atlantic Ocean. But they can still work with the two other countries. El Salvador will work with them. Probably some miscellaneous jobs like factories and manufacturing areas. I mean the whole thing. Why? Central America is between the Atlantic Ocean And the Pacific Ocean. They will work with regions, continents, or the entire world. Central America is the Center of the World anyway. The area is just too perfect. Even though the Pacific and Atlantic oceans are too vast, the countries and continents surrounding them are numerous and economically important.

With the Landbridge and the various industries that will work with it, the project will be very ambitious. It will be something that we can't just fathom. The economic and industrial strength will not only be dynamic and integrated with the world. It will just be formidable. It is a formidable region indeed. In the end, the economic development of those countries will accelerate once the industries are implemented. Too much. But it must be diverse. The Landbridge will be something, but other sectors must be implemented in addition to the manufacturing side. Let me guess. 10% Economic Growth is probably an understatement for the next ten years. Probably more than that. The impact will be tremendous. This effect will alleviate poverty. The impact will probably eradicate it.

That's because it will impact not only Central America but also the whole of Latin America, including the South American region. We have to integrate them economically, starting with Central America, of course.

Chapter 34
The Future Freshwater Of Central Asia, The Middle East, And Southern Russia

(It is a Lake, and It Might Save The Region and Probably the Whole World)

Caspian Sea – the largest lake in the world. It is so large that it accommodates a medium-sized country like Iran. It is so colossal; its waters can probably accommodate a region ten times its size. The area is so immense that combining the Middle East and Central Asia, including Pakistan and Southern Russia, is insufficient. Even the Xinjiang Region in China. But that's too far, but still reachable. The only problem is that it is a saltwater lake. Yup.

So, what should we do? The problem with the Caspian Sea is that it has no outlet. That means the water is not going out. I thought it was interconnected to the Black Sea, but it's not. The only way for water to go out is through evaporation. That means it keeps accumulating salt. So, if we can provide a way to get an outlet or a series of outlets.... The Black Sea and the biggest lake in the world, which is a saltwater lake, will become the biggest freshwater lake in the world—sorry, Lake Baikal. Number two is a good thing. The problem is that it may take a long time to remove the salt from the largest lake in the world. But no one cares. It will still work out as long as it turns fresh. But I am hoping for less than a decade.

We have found a way to make the Caspian Sea a freshwater lake, so let's do the more complex process. Spread fresh water to the necessities – the countries that badly needed it, such as Central Asia, the Middle East, Pakistan, probably Southern Russia, and the province of Xinjiang in China. Yes, all those countries. They need it badly for drinking, agriculture, and all other necessities. But can it be done? Of course, China did it with its South-To-North Water Diversion Project[38]. But this is more complex. But we cannot make it complicated; it should be simplified, not complex. So how do we do it? Well, there was a civilization that did it two thousand years ago – the Romans. They built one of the earliest water systems in the world, and for some reason, it is still there, complete with aqueducts, canals, and reservoirs using gravity. I believe some of them are still in use, but I might be mistaken.

So, here is my take. Build a water system so complex that it may need a reservoir or artificial lake to accommodate it and spread it further. Remember, we will turn desert or tundra countries into farmlands, forests, newly built cities, towns, and many more. of course, we need waterways coming from the Caspian for those necessities. Remember, many of these countries are rich countries and they do not care about expenses and probably build the largest and most complex water system in the world. That's because the desolated will be desolated no more. and they know the future will be immense and humongous that spending trillions of dollars on such projects will not be a necessity. That means no one cares.

But here is the issue. Once the water system is completed and starting to function, the water in the Caspian Sea will gradually dissipate. That's not good at all. So, what should be done? My answer to that is reverse water flow of wastewater. That means we have to build another complex system of reverse water flow of wastewater from the cities or places just near the Caspian Sea. Not those cities like Oman, Saudi Arabia, or Xinjiang Province if you know what I mean. I know it's complex, but it will not be complicated. It has to be simplified. Of course, it needs to be clean because we are going to use that recycled water in the Caspian Sea together with the freshwater from rivers to be used again. Remember, in the long run, forests will be created due to the abundance of water. Natural forests of course! That means rivers and streams will be created. It just needs to flow back to the Caspian Sea.

Now, we need a canal to provide access for ships from the Black Sea to the Caspian Sea – Eurasia Canal. It will probably be one of the longest large major canals in the world, but it will work. It just needs to accommodate the biggest ships in the world because countries such as Kazakhstan, Turkmenistan, Uzbekistan, and Iran, particularly the capital, which is Tehran, badly need them for shipping transport.

Then, there is the environmental problem. Once the saltwater is freshwater, the habitat in the Sea will eventually die. Everyone knows saltwater animals will not survive in freshwater habitats. But they have to endure. We can flush them out to the Black Sea, which is obvious, but the locals may not like it. That's because they want to preserve their habitat, of course, and we have no choice but to put it in a basin beside the Caspian Sea – Garabogazkol Basin. The animals should survive there, and the best part is that there's no outlet going out. So, it should be fine. However, the water has to be refreshed all the time, just like in an aquarium, but this is not an aquarium; it's a massive lake. So, it needs an outlet, but the salt must be preserved to ensure the animals' survival. I am not sure if we have that kind of technology yet.

There is no such thing as unreachable. Anything and everything can be reached. Such a process will always be a challenge. But humanity can achieve the unreachable, for humanity is immense in mind, heart, and hard work, and they are unstoppable. In ancient times, humankind could reach the unreachable, so our civilization can accomplish remarkably more. I don't want things to be complicated, but complexity is good enough. I hope everything else is simple because individuals and societies, too, like simplicity. We need to become "One." For everything and anything can be done if we are united... just like the Romans.

Chapter 35
South America And Its Weird Weather

(We have to control the Water Flow First.... Like a Game of Chess)

South America. Two major languages exist—Spanish and Portuguese. The words are easier to understand, yet the Portuguese agreed to have Brazil first, and the Spaniards will have the rest according to their agreement. The Spaniards have them all, including Mexico and even California. That's a significant advantage for Spaniards. How come the Portuguese agreed to that? It doesn't make any sense. However, the Spaniards were more influential back then. They destroyed the Aztecs, and I mean the civilization, just a few men. I guess they did know how to conquer the world back then.

Well, let's get back to South America. The real problem. The weather. Chile, which is in the Western part of South America, has the driest regions in the world. It is so dry that the sky is the clearest in the world. It is so clear that the most important telescopes in the world are mounted there. Now, they are building the two largest telescopes in the world, the Giant Magellan Telescope and the Extremely

Large Telescope (ELT), in a specific area: the Atacama Desert. This is one of the driest places in Chile, and I already told you that Chile has one of the driest regions. The reward will be significant because new wonders of the sky will be discovered once those two telescopes are completed.

And then, on the other side of Chile, which is the rest of South America, it is the wettest in the world. How did one side of South America become the driest in the world, and the other side is the wettest in the world? The town in Colombia is the wettest in the world because its annual rainfall exceeds 400 inches. It is the world record holder for the highest yearly rainfall capacity. It rains so strongly on that side that it creates the most powerful rivers in the world, and we all know the Amazon River is the most powerful in the world. It is one of the most dangerous too. During the rainy season, the river rises as high as 18 feet. Remember, the Amazon is not only a jungle, the largest in the world, but also a Basin, the Amazon Basin, which is the largest in the world as well. Of course, it has the largest network of rivers in the world, with the Amazon River itself being the leading actor.

The rest of South America, by the way, is Colombia, Argentina, Brazil, Venezuela, French Guyana, Guyana, Ecuador, Bolivia, Peru, Paraguay, and Trinidad and Tobago. Uruguay, Aruba, and Curacao. Of course, Chile is the main protagonist because its border, the Andes Mountains or Cordilleras, blocks the rain or turns it into banishment. That means the rain stays in the rest of South America. They said the Amazon is the lungs of the world. The problem is that the trees use the oxygen for what we call cellular aspiration[39]. That means the oxygen, which comes from the trees of the largest jungle in the world, never goes out of the world, and it always stays in that region. That's why we have to depend on the forests of Canada and Russia, because they provide more oxygen than the Amazon forest.

One thing about South America, especially the ones in the

Amazon Basin, is that it never loses its water. It is probably gaining more. It never gets out of the Andes Mountains, and that needs to be fixed. What are the solutions?

First and foremost, Chile needed its water. A new canal or a dozen canals need to be made from the Amazon Basin's Western border to the coast of Chile. The Atacama Desert needs to have green fields, as does the rest of the longest country in the world. The water from the Pantanal is already huge. That is good enough. The Pantanal is the largest swamp in the world, and it is also home to the largest rodent – the capybara. [40] However, there are others as well. Bolivia needs it as well. We have to divert water to them.

Argentina may need it as well. They have a desert in the south that lacks water. It is cold in the area, though, and still needs water. It's suitable for planting vegetables, vineyards, wheat, and many other crops. Pasture for their cattle as well. Brazil may need it in the southern part of the country, where there is water, but they might require more if they want to expand farms and pastures.

Central America may need it as well. They are going to build more canals besides the Panama Canal, and they may need these additional canals. The Panama Canal needs it too, as it is also lacking water. They may build two or more canals so that the eight to twenty-eight days of waiting for the ship to enter the canal will be shortened.

Northern Mexico may need it as well, for it is a deserted land. The population is still devastated by the drug cartel, and if they have different livelihoods like farming or ranching, then it may change their lifestyle or lives as well.

And of course, the Southwest United States, including Arizona, Utah, New Mexico, Southern California, Texas, Nevada, and many more, may need it. Those states lack water as well, and the South American Water can provide it to them. Again, rain in South America is recorded annually, and it is needed for various purposes.

Providing it to other countries outside South America will fix the water problem.

I forgot to mention that the Amazon River will provide water to the Sahara as well. It is in my Sahara Desert chapter. More information is provided in that particular chapter.

With the abundance of water in South America, the region must serve other continents. It is just too much water, and there is no choice. It has to spread throughout the continent of South America.

Chapter 36
Major Rivers In The United States Must Be Diverted Some Of Their Waters As Well

(Some rivers are partially drained)

We already talked about diverting water from South America. Then, we have to divert water from the rivers' abundant water to the rivers draining it. For example, the Mississippi River is fully plentiful in water. The problem is that there is too much water. Then, there is the Colorado River. It needed a lot of water. So, like the Rio Grande River in Texas. The state required it.

Las Vegas is one of the fastest-growing cities in the United States. It needed more water. Lake Mead cannot provide that water anymore[41]. Just too many people needed it. The Colorado River needs help, as does the Rio Grande River and other rivers that need assistance. It needs water from the Mississippi River or rivers and lakes that are abundantly supplied with water. This is the same issue with Salt Lake City. It is also one of the fastest growing cities in the United States, and only one major river is in it – the Colorado River and Lake Powell.

There is Salt Lake as well, but can we eliminate the salt? It's doable but it will take months or years. However, Salt Lake has to turn into a freshwater lake for the sake of water abundance.

So, who is going to do the job?

The US Army Corps of Engineers. They are the best military engineers in the world, and they can do the job. They have done spectacular civil mega-projects before, and they will do it again. They were remarkable during World War II which led to the demise of the Axis Powers which were the Germans Nazis and the Japanese Empire.

The West needs water and we can provide it to them. We need to thank the water diversion project that may lead to its success.

Chapter 37
Constant Rain Is Definitely What We Need And So As Climate Change

Farms... Forest Fires,,, Grassland Fires... This is the result of no rain for weeks or probably months. It is something that we cannot fathom with. For it causes destruction to villages, towns, or even cities. It is the result of drought and famine. Many more are attributed to constant drought such as lack of water, lack of food to eat, and animals dying. So, what is the best solution of these atrocities.

We need rain. We need constant raid to fight the drought. We need a way to do it. We need to find a way to clear this destruction. So, what can we do?

I found an idea. It works well with cloud seeding. The problem with cloud seeding is it needs clouds to work. If there are no clouds, it will not work. So, we need clouds to make it work. That means we have to make clouds. A lot of clouds. Also, remember that 90% of evaporation occurs in the ocean. That means we need evaporation because it will make clouds.

However, only few percentages of clouds are being evaporate

during summer. Plus, the hot air will not allow clouds to form. What should be done?

There is a Marine Cloud Brightening or MCB that I heard. Marine Cloud MCB involves spraying fine sea salt particles (aerosols) into the lower atmosphere over oceans. These particles act as cloud condensation nuclei (CCN) — tiny surfaces on which water vapor can condense to form cloud droplets. When there are more condensation nuclei, clouds form with more numerous but smaller droplets, which reflect more sunlight back into space (increasing albedo). Tend to last longer because smaller droplets evaporate more slowly. With this technique, we can cooler the planet and fight Climate Change.

When we combined MCB and Cloud Seeding, that will create rainfall. That will work well in area where there is drought and no clouds that is needed for cloud seeding. Because of that, we just did not fight drought. We fought Climate Change as well.

Just for information, Marine Cloud Brightening (MCB) is a proposed geoengineering technique designed to cool the planet by increasing the reflectivity (albedo) of marine clouds over the ocean. MCB tends to produce smaller droplets that are less likely to fall as rain — it stabilizes clouds. Cloud seeding, on the other hand, needs droplets that can grow large enough to fall. So, in some cases, MCB might suppress rain, while seeding tries to do the opposite.

So, this is what we need in places like California where there is not much rain to wet the forests. That's the reason forest fires are common in the area. What is more is it will help well in agriculture. And of course, drought. MCB can also fight Climate Change.

With that, we can finally find a way to fix drought and climate change. I just hope MCB will work well.

Chapter 38
Airplanes

(It Must be Bigger, More Efficient, and Sustainable)

With the demise of the Airbus A380 and Boeing 747-8 and the demise of the Concorde Passenger Jet a decade ago, innovation in airplanes has become a struggle. Airlines decided to buy smaller but efficient airplanes. Did they make a mistake? Yes. Big mistake. Passenger demand is still great and the likes of the Airbus A380 are full of them. I am not sure about Boeing, but more and more passengers are flying or traveling. We needed bigger planes. We need fast ones as well. Passenger demand has been at an all-time high since before the pandemic and it is going back to normal.

We designed and built bigger passenger planes because they are necessary. We designed and built faster planes because they are necessary. What we need is patience. That is because I believe the airline companies will still order those planes and the demand for them will be higher in years to come.

We have to rebuild them. We have to redesign them to make

them more efficient and sustainable. We probably have to make it electric. We can. However, hybrid is good enough. And it does not need to emit a lot of smoke. The lesser, the better, of course.

Passenger Planes have been innovative since the Pan Am days. It is now gone. However, it is still in our history. What we need now is more passenger comfort. I know the economy class is smaller, but we have to make it bigger and more comfortable. It is the only way. Passengers need to choose which one to buy as well. That means businesses inside the airplane like restaurants, souvenir stores, massage places, and many more.

That's why we need bigger airplanes. The reason is the accommodation. We needed it badly.

Knowledge and power
Such qualities, such talents
Yet love can conquer it all
For it can conquer the world

Chapter 39
Bring Back Tv Shows That Has Meaning And Importance

(Too many Reality TVs)

I am probably the biggest fan of the Cosmos TV show hosted by Neil DeGrasse Tyson. It was once hosted by Carl Sagan, a very popular icon who died at an early age. Cosmos was a learning show. It developed you. It makes you think deeper and makes you imagine more. It provides you with the history of the past and future. It teaches physics, past and ancient history, politics, global affairs, warfare, and many more. It is a complete subject.

It is the biggest reason why I am writing this book.

Cosmos opened my mind. It made me think more deeply and become more curious about this world. It made me feel that the world can be a better place. And it will be a better place if you follow my book. I hope there is a movement off of this book. People will go for a revolution because they want to be different. They want to be the ancient Greeks. They want to be like the ancient Chinese. They want to be innovators. They want a revolution for a better world. A world where everything is possible and doable, and they will do it freely while money flows into their pockets for innovation, providing ideas, and knowledge. I want people to change for the better.

I want people to change for the better.

The world is getting better anyway. We are getting better. Poverty is alleviating and we become innovators. But I want more. I am not satisfied.

I am not satisfied.

So, what do I want? I want TV shows like Cosmos every day for the rest of people's lives. Not Shark Week. Nor those Reality TV shows. I am not trying to insult them. I want something better. Can we implement it? Yes. We can. We just have to do it. We have to believe in it. We have to do it. This is for our children anyway. This is for their children, too. And for the next generation to come. I want Cosmos every day or every hour of every day. That is too much, but I want one. It will make your mind bigger anyway. And that is the goal. To make your mind bigger and better. To help you think and imagine better and bigger, I play chess too. It will make my mind think strategically. It will also improve my mental well-being. And I am good at that too. I was a High School Chess Tournament Champion thanks to that game.

I just want all of you to know that TV shows influence us. We have to make sure we are watching for the right influence. For the betterment of the world.

Thank you.

I may be weak, but my heart is strong
I may be stupid, but my mind is just in dormancy
Your love is the key for me to rise and become great

Chapter 40
Hong Kong Must Not Demise

(Still the Gold of Asia)

Hong Kong.... Once the darling of Asia. The Great Arch to China, and most of all, the freest city in the world. It's the reason why China wants to turn into capitalism. And it did. If there is Hollywood, there is Hong Kong. A great city indeed. It was the greatest city where people flocked to live even if the standard of living was the highest in the world. They continue to flock. But it stops.... That's because that was in the 20th. Century. It's different now.

Hong Kong was once the opening gateway to China. What China manufactures, which was in the Pearl River Delta region, always moves to Hong Kong for shipping around the world. China was a strict communist country back then. But not Hong Kong. And Hong Kong profited way too much because of it.

But something changed. Something became indifference. China finally opened up and so as its cities, ports, factories, and many more.

Pearl River Delta became the factory capital of China even though China is the factory capital in the world.

But not Hong Kong. It did not commit to change. It did not commit to progress. That's because its people were content enough with its way of life and standard of living. Hong Kong did not commit to reform. As I mentioned in the last chapter, Shenzhen became the fastest-growing city in the history of mankind. Sadly, Hong Kong became dormant, and it stayed dormant. China is now the second biggest economy in the world. It has four cities that have a bigger GDP than Hong Kong. And that is Shanghai, Beijing, Guangzhou, and lastly and dreadfully, Shenzhen. Do you know why I said that? Just like what I said in the last chapter, Shenzhen was a fishing village more than 30 years ago. And now, the new capital of Pearl River Delta – which is now the biggest metropolitan area in the whole world. Factories are diminishing but being replaced by Technology companies. And there are a lot. A lot of start-ups too. Even Unicorn Start-ups.

But not Hong Kong. For some reason, its dormancy continues, and it is becoming more rampant. It is still beautiful, and investments are abundant, but it needs a lot of changes. Why? That's because we do not want Hong Kong to become a dormant city. It must act or react. It must not be allowed to be destroyed and turned into a nightmare city or I can say, not in a disrespectful way, a sinful city. A city for the underground world, full of organized crimes, notorious gangs, and a broken mirror of destruction.

So, what Hong Kong must do? What should it transpire? What should be the destiny?

First, Hong Kong is just too expensive. The city is not in demand anymore. That means it's not essential for other countries anymore, especially China. Hong Kong must learn that Shenzhen and Guangzhou are the faces of China in the Pearl River Delta now. Not Hong Kong. That means it cannot be too expensive anymore but still remains competitive. Is it hard? Of course. But it's doable. That means it needs to make some sacrifices to bring it back to a competitive and balanced city again.

Hong Kong is still the second center of Southeast Asia. Of course, Singapore is the center of Southeast Asia. But Hong Kong? That's in China. Not in Southeast Asia. However, the countries of Southeast Asia remain committed to Hong Kong. And Hong Kong has to stay committed not only to Southeast Asia but to the entire humongous world. That means Hong Kong has to be more accessible than being the freest city in the world. That's absurd, but it must be done. It doesn't have technology start-ups that can compete against the rest of the world. It doesn't have financial services that can compete against the rest of the world. It was once had, and that is HSBC (Hong Kong Shanghai Banking Corporation). The problem is the company's headquarters are now in London. The famous building in HK is still there though but it's not the HQ anymore.

Hong Kong needs to welcome international organizations. For example, the United Nations. It has headquarters in New York City of course. It also has HQs in Europe and Africa. But for some reason, there is nothing in Asia specifically East Asia. But Asia is a very important region now. International Organizations like the United Nations must commit to placing another HQ in Asia. And that's how Hong Kong needs to be on the frontline. That's because Hong Kong's geography is very important. It is located almost between the center of East Asia and Southeast Asia. That means dignitaries and leaders can fly to Hong Kong with ease.

Hong Kong must allow to commit innovation. It has the best colleges and universities in the world but the technology start-ups or even unicorns, are almost non-existent. Hong Kong must do that.

Then, I found another issue with Hong Kong that must be solved. And that is the Real Estate market. The most expensive real estate market in the world. The problem with Hong Kong is it needed more space to build. It does. The problem they need to do it. They need to start building a new urban area somewhere in the Special Administrative Region. The region still has no urban areas. And in Hong Kong, there are still a lot of them. They are mostly concentrated in the Eastern part of Hong Kong. The only problem is they are mostly reservoirs. Well, 70% of the water supply in Hong Kong is from the Mainland mainly in Guangdong Province. So, we don't need to depend on the local reservoir. Actually, we can still use them. If that's the case we can live in an urban area with a water reservoir. We just have to make sure that the reservoir is clean the way it is.

Some countries did that as well such as Singapore. They even recycled their water. So why not Hong Kong? They can build new buildings over there, and the construction will be cheap. Hong Kong's price is low. It's not even as far as Kowloon or Hong Kong Island, which is probably a 30 to 60-minute drive in rush hour. The good thing is that the rent, or the mortgage, will be cheap.

And that's the key. Everything in that particular area will be cheap. It has to be cheap. For that is the reason for the city's would-be demise. We have to fight the actual reason for our competitiveness.

For that, we can achieve re-flourishment— a reincarnation and a second chance.

Hong Kong still has a lot of potential. We can put new start-ups in those new urban areas. We can put the Non-profit Organizations there, specifically Non-Government Organizations or NGO's. And the whole of Asia needs them. For example. The United Nations itself has headquarters in New York City, two HQs somewhere in Europe, and then there is one HQ in Nairobi, Kenya. So why are there none in Asia, which has probably 40% to 60% of the world's population? We must build a United Nations Headquarters in Asia, and it will probably be in Hong Kong. As well as other prominent and influential organizations, and probably Think Tanks. Again, the location of Hong Kong is just perfect. It's near every country in the

Asian region. It's ideal for leaders and dignitaries to go there because it's near and one of the most accessible cities in the world.

Hong Kong cannot stay behind. While Shenzhen and Guangzhou are flourishing, HK is becoming non-existent. It must not allow itself to be overtaken. No worries. HK still has a high GDP per capita. But remember, it's not about that. It's all about the city itself. That means GDP is more important than GDP per capita. That's because GDP represents the entire city or country, while GDP per capita represents an individual person. Of course, individualism works, but together we will stand.

Currently, the trajectory of Hong Kong is murky. That's why I am giving such clarity. Is this the only way? Of course, there are more. The one thing I want to provide is a light. For I am a Luminary, and I guess I have to carry God's work – to carry the light to its right destination. The world needs a light. It is not me of course. But it's the society. A golden society. For they will tend to carry the light into the Golden Age. The Age of Renaissance. Believe me. They can. They are still young you know. But it's not only Hong Kong. It's everybody. Of course, it's the young and the old. And again, of course.

Chapter 41
Integration Of The Middle East

(The Middle Must Be Combined for the World Can't Be Combined Without It... Geographically, of Course)

Chaos... Chaos... and more Chaos... That's what we see in the Middle East. A chaotic world. A world full of war and conflict. Never-ending wars. Never-ending catastrophe. It is somewhat of a regular occurrence now. It is a never-ending regional event.

We presumed it was always like that. A region that has no peaceful means of solving things. People thought Middle Easterners were a bunch of uneducated, chaotic, and merciless with no peaceful solutions. Is that always the region? Of course not.

The Middle East was once the most prosperous and innovative with the most intellectual people in the world. Remember, Arabic Numerals are Arabic, of course! Baghdad was the Golden City. And of course, back then, was the center of the world. While Europe was in the Dark Age, The Middle East was in the Renaissance Age. Of course, I can't say it's like that but it's almost like that. That's because they never knew they were in the Renaissance Age. But were they chaotic? Of course. They have no choice. The world is in chaos at that time, and they have to adapt to such an environment. But that was the Middle East back then. I want the Middle East to be like I want it to be intended to be. That's because they are learning, and they will not stop. They just need a guide. A luminary. That's not me of course. It is the society. And society thirsts for such change. For such reform. For a new progress.

Middle East Integration is possible, doable, and of course, can be done right away. The problem is it is just chaotic. But chaos can dissipate and diminish. It just needs to find a way to do it. It already started. Iran and Saudi Arabia finally have their unwanted intentions and rivalry solved. You have to thank everybody for it but China was the key player. Of course, the rest of the world accepted it. It's the only way.

Iraq, which is a country that separates the country of Saudi Arabia and Iran, is finally peaceful too. Another key of course. Syria is still an issue for Israel and the Palestinians. But as I mentioned again, it will dissipate and diminish. What it needs is to provide eternal peace, which, of course, is just a liberal idea. But of course, the world wants that. So, it's not a liberal idea of course. It's just a liberal ideology. But I am not a liberal, conservative, communist, or combined. I want the right solution to be provided. I want the best way to follow my beliefs. I believe that the Middle East can be resolved, which means the impossible issue can be solved.

So, what else does the Middle East need? Turkey's Economy is so remarkable that its economy is outpacing the rest of Europe. Pakistan's Economy is also improving moderately, but the China-Pakistan Economic Corridor promises to change and reform the country in a significant way. And no one is stopping it. Remember, Pakistan has more than 200 million in population. That's a lot of consumers.... In the future of course. They need to make money. The

Gulf State Countries include Kuwait, Bahrain, Qatar, United Arab Emirates, Saudi Arabia, Oman, and unfortunately, Yemen. But it can work out. That means the issue in Yemen can be solved.

So, what else do the Middle Eastern countries need? The basic economic driving tools, of course, are Transportation, Communications, Social Reforms, Progressive Reforms, Manufacturing, Investments, and many more. The most important thing is to implement them first.

Of course, roads need to be built to integrate the economies – internally (cities) and externally (countries). That means Turkey, Iran, and Pakistan need to interconnect with those roads. Not just any other roads, but Expressways, of course. Why did I mention it? That's because those countries are separating Asia and Europe, and they must interconnect. So, as Iraq, Syria, and many others. But the key is the integration of road networks. Communication is vital too, but that's easy. It just needs to be secured. Social Reform is possible because Saudi Arabia and the United Arab Emirates are leading the way. Women are now equal to men, and the same goes for expats and other foreigners, of course.

Progress is being made, and they are doing well. We have to thank Dubai for that. A remarkable city and the center of the world. One of the centers, of course!

Then the key is the economy. The most important aspect is to improve society. The integrated economy must be achieved, of course, as should the local economy. The Middle East has a good coastline as well, with the Red Sea and Suez Canal being one of the most important shipping lanes in the world. Of course, there is an issue right now, and that must be solved.

Then the Israel-Palestinian conflict. I already provided such a chapter, and I hope they will find a way to solve it.

Lebanon is another problem, but it can also be solved, as well as Syria, and Iran, which is currently run by Shiites, who have different ideologies, which is an ideology that will work well, but it doesn't provide flexibility. The key is for the Ayatollah to provide an economic means to solve that issue. That means it needs to open, and I hope he will find a way to work with the West to solve the sanctions they imposed on the country of Iran. But I believe the government has the backing of the rest of the world. They need to prove they can comply with those sanctions.

The Middle East is not an Oil-Rich region anymore. I mean, it's of course an Oil-Rich Region. But it's different now. It's a region that can integrate and function as one, combining diverse industrial sectors with a progressive society. They can't depend on oil. Of course, it's a big industry, but they must allow us to find another means and diversify. This is what countries can do, and groups of countries can save by any means because they work as one and cannot be broken. They are a remarkable group of countries after all. And they will prove it.

And then I assessed the land that borders the Gaza Strip, Egypt, and Israel. We know the Suez Canal can accommodate ships crossing it every eight hours or more. I believe it needed another one to help it. I propose a Landbridge. That is the strip from the Israel-Egypt border going to the Gaza Strip. That will accommodate the three countries. Can I say three? Can I include Palestine with it? Can you permit me to do it?

All I want is for the state of Palestine to have a direction. They need a direction. They can have Israel engage them with all means, but all they want is a future. All they want is a rejuvenation. All they want is a promise. All they want is justice, not punishment.

I am proposing a Landbridge. It needs the biggest container ports in the world to accommodate it, and I believe the Gaza Strip can accommodate it. Imagine Container ports in the Gaza Strip, people working to deliver those containers from one end to another. Some have to work overtime because the demand is just there. So, is that the reward? No, they want more. Of course, I provided the so-called reverse domino process. The process involves accumulating demand from all sectors and ensuring it is accommodated for the people. They need more restaurants, schools, universities, SMEs, and many more. They have to do it while the wealth is being provided at a low cost. That's because once it becomes more expensive, it will be hard to offer more financial incentives, as investments will dwindle, and consumers may opt for a cheaper accommodation. Probably, in another city besides Gaza. So, what I meant is take the opportunity while it is early. Gaza needs everything it requires to achieve what it calls the Golden Age. The Renaissance that it wants to have. A new Palestine. Where people don't have Israel but hate what happened to them and everyone else, should I provide it to the War History? No. That's because another country will help. That is everybody in the Middle East. That is everybody in the world.

Chapter 42
Turn Sahara Into Atlantropa

(In fact, let's turn any Desert into an Oasis.)

Atlantropa[42]. Once was a dream. A dream that would come true. But it's just too fictitious. Too impossible. In fact, its concepts will not work at all. It was once devised to transform the Sahara Desert and parts of the Mediterranean Sea into fertile land. The problem is that it's too conceptualized and won't work because of the technology in the 1920s. It evaporated during World War II due to the hardships of its creators in Europe. It was not a concept anymore. Just a fictitious history that will not turn into reality, for it completely disappears into people's minds.

But we have to bring it back. It has to come back. It needs to come back. For the desert is big and has its grandeur. Of course, it has a lot of opportunities. The desert can bring an almost unlimited need for resources. It can bring bountiful resources. We have to build one. We have no choice but to make such fertile soil because it is badly needed by the billions of consumers there – Europe and Africa.

So, what now? We have to build the biggest fertile land, probably in the world. That fertile land can turn into farmlands, pastures, towns, and cities. It will be an enormous opportunity for the Central African Region, which is one of the poorest in the world. It will be an enormous opportunity for North Africa. In fact, it will give everyone in the world confidence and security in areas such as food, jobs, economic opportunities, and more. Just build one and they will come. Actually, they are there. We have to go and aid. We have to carve. We have to sculpt the Sahara Desert together with the rest of the world.

So how do we turn a desert into fertile soil? Of course, it's the water. Water is the most critical resource for all beings including humanity themselves. Without water, the land will not be usable. Without water, only a few people or animals can survive. In fact, it's a harsher world for the survivors.

So, water is the key, and we have no choice but to find it.

Thank God we found it. It is beneath the desert. It's called groundwater. And thank God, the Sahara has the biggest groundwater systems in the world. So big, it may accommodate the whole desert. It's so big, it can turn into fertile soil. But here is the problem. The source of the groundwater is rain. And it doesn't rain all the time in the Sahara. The key is to find a constant supply of water. The groundwater can probably supply the Sahara for months or even a few years, but in the end, it will be gone. The rain will not allow it to resupply with more water.

So, what is the following process? Central and Southern Africa have one of the most abundant water resources in the world. It would be awesome if we could find a way to distribute those bountiful waters from going South to possibly going North, which is in the Sahara Desert. So, what is the process for that? It's probably not easy, but we have to plant more trees and fight the erosion thatis happening on the border of the desert. And of course, distribute it in all parts of the Sahara so that it will finally turn into fertile land or green fields with farmlands and pastures. The key is that once the soil becomes fertile, we need to start planting trees in it to turn it into a forest or jungle. It's possible. A couple did it in South America, and their land is now a jungle with all those birds and animals. It was arable land a few years ago. The problem is that it is a desert, and we have to fight that desert and turn it into fertile soil, and turn that fertile soil into a jungle. That's because if more jungles are made, then more water resources are also made. The key is to distribute that water into the Sahara. Remember, a couple did it. A million people can do it too. In fact, a billion organized people can make such a humongous project simple. By the way, that humungous project is probably an understatement. I mean, yeah, it is.

And of course, we can get fresh water from the mouth of the longest river in the world – The Amazon River. We will just have to get it from the river, send it to Africa, and spread it into the Sahara. Remember, the Amazon River has so much water, it causes devastation to the places near it because of that. The Amazon will accommodate the Sahara with its fresh water. It is going to be abundant, and we just need several pipe tubes to transport the Amazon water from South America to Africa.

I believe Northeast Africa especially Somalia needs fresh water as well. We can accommodate that by providing water from Lake Victoria to that area. It will work well because Lake Victoria is a humungous lake and one of the biggest freshwater lakes in the world.

Besides the Amazon water, East Africa can be accommodated by the Congo River as well – one of the most powerful rivers in the world. Second only to the Amazon River. It can provide such water too just like the Amazon River.

So, we have to build artificial rivers going north with artificial lakes and other stuff to distribute it to the desert. Remember, it can be done. But it is humongous. But do the countries in Africa care?

Of course, they do. I mean it's the biggest opportunity and it will turn those countries into rich countries if done the right way.

We just have to tell them it is possible. That's because we can carve the world, we can also turn it miracle into reality. For we are the new generation of humans. Humans who can carve the world. Humans who can do miracles. Humans will be superhuman generation models for the next generations to come. We will turn the Sahara into Atlantropa. We will build towns and cities with it. We will build a new civilization. For the Sahara is now Atlantropa. And Atlantropa is the model of our generation – the Carvers of the world.

Of course, we can do other deserts as well. The only problem is it has to be another chapter. Or no need for another chapter. The Sahara model is good enough. It is a chapter for Atlantropa of course. And Australia and Western America have opportunities as well. We just have to carve them too. Our generation, of course, is the carvers and sculptors of the world.

I appreciate the vision made by German Architect Herman Sorgel for his Atlantropa Project. I know the project died with you. But in the end, we want to thank you for your immortality. Atlantropa will always be immortal to you.

Chapter 43
Jefferson State... The New Golden State

(It Ain't The Best Place But Let's Make It The Best On the West Coast of the United States)

Another vision. Another ambitious indecency. Another imagination is waiting to deter the enormous length of the vanished state. The state that doesn't exist. The state that went off the history. The state that was never a state. A well-being that was forsaken by its ambitious indecency. Jefferson State. Once a proposal to be the 51st state in the United States is no more, but a mindset of the vanished citizens who want to be visible and be it as their own[43], yet it is gone. And it will not come back. For the state is an ambitious one. The country of the United States may carve it as one of the most ambitious in the world.

Let's do that. Let the United States carve the Jefferson State. Let's make it visible. It's an ambitious indecency indeed.

California was the biggest economy of all the states. But it became more expensive. It became one of the most expensive real estates not only in the United States but also in the rest of the world. Yup, 1 million US dollars on average. Such an average person can't fathom normal people with a decent life. They can't afford it. But they won't leave it. They love California. They love the so-called Golden State Habitat. A habitat that is so unique that the whole world thought it was the entire United States. Baywatch may have secured it. Even though the series was bad, the rest of the world watched it. Yes, immigrants want to stay there. Millions of them. They don't care about the expenses. Because for them, the habitat seems to be decent. The habitat appears welcoming. The habitat seems to be the environment they can deal with. So, as the locals. The Californians, of course!

Yet, it is becoming dangerous to live financially there. But people still want to stay. Many become bombs. Living in the streets. I heard those bombs are not really poor. I listened to some lawyers. I heard some are doctors. I heard people are financially decent. But they can't afford the real estate.

So, what is the best solution to all this chaos?

Yeah. Let's build a new city. Actually, it already existed. Let's make an invisible state that will integrate other towns, for we have found a land that can provide that. We have to take advantage of that because it has diversity in land, resources, and many more. California has no water. I stayed there for three months, and it never rained even once. But it has to rain. And it does in other seasons, but not in Summer. But it's a dwindling rain. It is never a rainy day that they can depend on. But that place seems to be decent. Actually, it has the most rainfall of all of California. The problem is that it is in the land of nowhere. Yes, there are towns and cities. But they are small and not decent.

So, let's assess it. Jefferson State, a state that doesn't exist, will make this land the best land because water is abundant. Water is reliable. That's the best resource that anyone can rely on. Yet, there are more. There are towns and cities. Again, they are small but decent. That means room to grow. More to put new people in, more to put companies in. Probably, the next Silicon Valley. Probably, just another technology mecca. That's because real Silicon Valley is just around the corner. So as everyone else. And of course, the real estate is cheap. Really cheap. Again, there is room to grow.

Then, there are the vineyards. A wine country. It is just nearby. It is reachable—no need to go anywhere else. Wine is accommodating. The wine will be cheap as well, for everything is cheap. Everything will be affordable for everyone.

And of course, the one that I can't fathom. The integration of the towns and cities is underway. In a different way. More like diversifying the cities and integrating them into those cities. This is once formulated in the City of Dubai. But it was never implemented. But they may one day. For example, there is a technology mecca. There is a casino city. That's because Reno is pretty far. And of course, Las Vegas is much farther. There are the vineyards. There is probably a medical city. A Science City. An education city. A leisure city, and of course, a National Park, which makes California famous. But there is nothing there. Well, let's make one. But to tell you the truth. All of the surroundings seem like National Parks themselves. So, it's going to be a lot of National Parks. Ambitious indecency indeed!

And then the rivers. It has the second-longest river in California. But there are other rivers as well. But it's not about how many rivers. It's about what's in it. Of course, rafting and kayaking will be a good choice. But the best option is freshwater fishing. That's because it has the best freshwater fish in California. Actually, we can say it's the unique one in California. But it will be good. It will be something.

The only issue is the interstate highway to get there. There is nothing. Well, let's carve one in the mountains. Let's make not one but probably a lot more. The important thing is we need to use them to integrate all those cities. One thing about California is that it has the world's best urban and state planners. They have to prove it in this Jefferson State. We must provide the best urban, suburban, and rural planning. And we have Silicon Valley to assess that. We will use vehicles, of course. But the future will be autonomous vehicles. The vehicle that we don't need to drive is available. They said if all cars were like that, then traffic lights would be nonexistent. That's because AUVs will calculate the intersection for other vehicles before they cross that intersection. And, of course, the goal is not to intersect. It will work. We have to make sure that it is reliable. That's because the technology is still in its infancy. But it will be perfected many years or decades from now, and vehicles will no longer intersect.

So, what am I saying? What is my goal?

I don't want to create just a Jefferson State. I want to make a state so advanced that it will create a society of the same level. It has to use a lot of advanced technologies. It has to create something non-existent. It has to make a technology that will not follow the fundamentals of physics. Yeah, it is just too much. It's ambitious indecency indeed!

Ultimately, I want it to start in Jefferson State—a state that doesn't exist—and spread to the rest of the United States and the world. We need the state to exist. It will. We have to carve it.

But we have to let them know. You know. "THEM." The people with ambitions and visions.

Chapter 44
India Must Endeavor, And So Must The Rest Of South Asia

(Once they have their pride. They will have it again.)

They were once one of the most formidable countries in the world. Their economic power is colossal. They were the most dominant economies in the world. Their military might deterred Alexander the Great from conquering the land. They once had one of the best educational systems in the world. The education system is so advanced that it created the best ancient university in the world – Nalanda[44]. With ancient education comes the great minds of the world. The knowledge they possessed was so unimaginable that it took another thousand years for the West to acquire some understanding of it. Examples include gravity, ancient astronomy, ancient medicine, and various mathematical concepts such as the zero and number system, the decimal system, the value of pi, algebra, trigonometry, calculus, and many more.

The Indian Civilization fell into further decline and devastation due to poor leadership by its successive rulers. The situation became more complicated when Islam was introduced in Northern India. This event saw the end of indigenous empires in India, and the civilization was never the same.

However, colonization caused many extremities in Indian society, and it took the country several decades, probably a century, to regain its near-formidable power. It was different back then. Such extreme behavior and actions occur in colonized countries. India suffered the most because it had one of the biggest economies in the world, and the colonizers knew it must be contained to ensure control. It is unworthy, but such an occurrence will never happen again.

But India is now in a different midst. It is becoming unrelenting, a country unprovoked by a population hungry for more wealth, power, and influence. The country's GDP is growing so fast that it will take just a decade to become the third-biggest economy in the world. And it will grow quicker and more formidable.

India is at the forefront of the services industry. It is a prowess in Information or Software Technology. But the best thing about it is that Indian people are more confident, energetic, forward-leaning, and, best of all, hard-working.

But something is lacking. Something in India must be enhanced or curtailed to experience economic growth, integration, and dominance. Such economic power can only be eclipsed by one country: China, which experienced the fastest economic growth in human history. India may follow that or even break its record.

So, India is doing what it must be doing. First, it prioritizes infrastructure projects. It needs highways, rail tracks, public buildings, hospitals, public parks, and many others. It requires those to function as vessels for the country, for the hearts, which are the towns and cities, must function to provide commercial and economic integrity.

I mentioned the Gestalt Law in Africa, which includes the laws of nearness, accommodations, simplicity, easy reach, and accessibility.

Africa, as a continent, can relate to or even encompass that. India is ideal for the cities' proximity to this methodology. The country has dozens of cities with populations ranging from millions to tens of millions. And they are all near to each other.

So, what is needed? India already has expressways, but there is a need for innovation in its railway system. The new concept I propose is Maglev or Magnetic Levitation, a cutting-edge technology that offers faster and more comfortable travel. With Maglev, passengers can travel at speeds exceeding 300 mph, equivalent to 400 kph, without feeling any discomfort. This technology has the potential to revolutionize India's transportation system, making travel faster and more efficient. Of course, Maglev is still expensive, so High Speed Trains on Rail Tracks will be more efficient and cost less.

I noticed the Indian Buildings are big and wide. That is not good. Remember, India is too small for 1.4 billion people. China can accommodate that, but India needs more space than ever. Besides, some elites may need a large amount of land for their new home. That will not work. India must build high-rise or skyscrapers. Remember, local farms are still required in India. And 1.4 billion Indians needed a lot of food every day. We can accommodate high-rise urban farms, but traditional farms are still better and more flexible. Urban farms may work in the future, but we still need traditional farms to provide food and satisfy our appetites. Unless Tibet or Pakistan does the job? They are not local farms anyway. And the food is highly demanding. In the future, once their lands are ready for farms with new technologies and a new environment suitable for farming, they will likely develop them.

The country needs to innovate as well. It has many technology companies, but it must do more. Technology companies must diversify and integrate to ensure that innovation and collaboration are fast and successful. It must ensure that creating businesses in technology and all others will be smooth without government and competitors' intervention. Creating such companies must take only days or weeks, not months or even years.

Of course, education is a priority as well. It is one of the best in the world and one of the largest, too. It has many college graduates, but many of them are unemployed.

The government must tackle this issue. It needs to find ways to provide jobs internally, not externally. That means it must refrain from exporting people to find employment in other countries. It must prioritize private business creations, technological innovations, and manufacturing, while also encouraging investments, mergers and acquisitions, and government interventions in the market and the economy.

The creation of the middle and upper classes is also a priority. To do that, jobs for the masses must be created. And those jobs are in manufacturing. India must make and prioritize manufacturing businesses. The businesses that will generate hundreds of millions of jobs. It must do that with keenness and precision. For manufacturing needs a lot of different entities to provide a manufacturing environment. It must build roads to accommodate supply trucks. It must create more seaports and airports for transportation of goods and services. It must prioritize logistics for goods, and materials must be distributed well. Most important of all are job training and retraining.

India also has to prioritize its geography. It is the biggest country in South Asia. It is the heart and core of the region. The government dominates the Indian Ocean itself. And that is a priority in itself. It is also in the middle of Asia. That means it can intertwine with the Middle Eastern region, the East Asian region, and the Southeast Asian region. Even East Africa and Europe will be of utmost importance. It has to take advantage of its nearly perfect geography. That means more aircraft and ship vessels are needed to do the job.

If India wants to see the furthest stars, galaxies, and the whole universe, it does not need to go to Chile to make the largest telescopes in the world. Tibet should be good enough. It has one of the clearest skies in the world. It's probably clearer than the Atacama Desert in Chile. China may use it, too, for their telescopes. It's their country, anyway. It is just too far and isolated. But roads can be built to get to the right place in Tibet.

So, what else, Winter season? Of course, India doesn't have it. But the Himalayas have it, especially the country of Nepal. The Himalayas may accommodate the biggest ski resorts in the world, with slopes that will take 30 minutes to an hour to get to the end. That will be tiring just for one slope, but who cares? Plus, it's a green slope. Double Black Diamond might be deadly, but that's good enough for thrill seekers or extreme experts. Again, it's lethal. Skiers from around the world will go to the Himalayas for that. We have to build those ski resorts. Remember, the Himalayas have 30 to 40 of the highest mountains in the world, including the highest, Mt. Everest. Imagine building ski resorts on the highest mountains in the world. That will be deadly, and profits will be soaring because it's Winter in the Himalayas. We have to make sure that the Avalanches will never occur. We do know how to do that. Safety is always the priority anyway. Of course, hotel resorts are poorly needed, as is everything else. Will Nepal make a lot of profit? So, India occupies the rest of the Himalayas. It's probably China, but they are on the other side, which is pretty far. Who knows?

The Indian people must understand the concepts of cooperation and competition. Cooperation is needed to ensure networking integration, foster collaboration, and ensure that no one is left behind. Everyone will prioritize and perform well in their jobs. It must prioritize and provide a competitive environment to prioritize the best or the dominant.

India must prioritize the military as well. There are still issues with China and Pakistan, which must be set aside. Pakistan is probably softening, but China is an issue that must be fixed. They must prepare the military and provide better weapons and equipment to ensure peace and stability. Pakistan might be an issue in the future, for terrorism is coming back. But India knows how to protect its borders, and it must defend its borders to deter terrorists.

Banking and Investments must also be prioritized. India is a country full of opportunity. Foreign investments will increase yearly as more investors invest in India's economy. The investments will create more jobs, communities, and economic growth.

Populism must be eradicated. It will devastate the economy, society, and, most importantly, the leadership. Narendra Modi must know how to unite the government. He will always be the key to it. Populism must be stopped. Minorities must be prioritized.

There are many opportunities that India can seize. These are the only ones I can relate to. This is simple to implement yet hard, for bureaucracy will lead to nothingness. Narendra Modi needs to have complete control of the situation. He does not need to be a dictator. He needs to unite the people. For the people who think they have a future, they will not relent in accomplishing it. The Indian people are unstoppable and always formidable.

My heart is unwealthy
for I do not have you as my opportunity
Be one, and my heart will find astonishment

Chapter 45
A Must Urbanized Rural Areas And Farm Productivity Innovation

(They are still the Key to Economic Growth, and the Economic Dominance for their Importance has to be prioritized)

Rural areas often have neighbors who are far from each other. You occasionally drive to your neighbor's house to talk to them. That's the experience of the Rural United States or Rural America—the experience of almost isolation. However, I cannot really say that isolation is the best word because I am not a rural person. However, if there are any emergencies, your neighbors might not see you, especially if your house is on fire, someone has broken into your front door, or there has been a burglary. The question is, do farmers need their houses on the farm where they are far away from their neighbors? Or can they be near their neighbors but a bit far from their farms? I believe the latter will be better.

The farmer's house doesn't need to be at the farm thanks to the latest technologies such as Remote Monitoring Systems, Automated Machinery, Smart Irrigation and Fertilization, Cloud-based Farm

Management Software, and Remote Livestock Management. Let us discuss these technologies, as they will benefit the farmers significantly. This is according to ChatGPT.

First is the Remote Monitoring System. It is a system that uses drones and satellite imagery to provide real-time data on crop health, soil condition, and irrigation needs. Internet of Things (IoT) sensors are also being used to monitor factors like soil moisture, temperature, and livestock health. It will send any updates to the farmer's smartphone or computer.

Second is the Automated Machinery. Autonomous tractors and harvesters can operate with minimal human intervention and be controlled remotely. Robotics can also be used for tasks like planting, weeding, and harvesting, reducing the need for constant on-site presence.

Third is the Smart Irrigation and Fertilization. Systems can be programmed to adjust water and nutrient delivery based on real-time data, ensuring efficiency and reducing manual labor.

Fourth is the Cloud-Based Farm Management Software. Farmers can track inventory, manage finances, and plan crop rotations from anywhere with internet access.

The fifth and the last one is Remote Livestock Management. GPS Collars and biometric sensors on animals allow farms to monitor their location, health, and activity remotely.

Having a remote farm is a significant benefit to the farmer and their family. They will have better living conditions because they can access better schools, healthcare, and amenities in urban or suburban areas. They will benefit from having their kids receive a better education, easy access to hospitals or health centers, and the ability to take advantage of restaurants, shopping centers, and more due to its proximity. They will also be close to the proper authorities, like the police, fire department, and other authorities. It will be a significant advantage for the farmers and their families. It can also improve their work-life balance. It will reduce stress from physical demands and the isolation of farm life.

All in all, a remote farming system is an advantage that will not only benefit the farmers and their families but also allow them to thrive significantly and reduce pressure. Farm life will never be the same again. It will be easy and better.

Chapter 46
The Worthiness Of The Virus

(The Pandemic That Decimates Us All)

COVID-19. The virus started in a common market. They ended up as the ones who decimated us all. A virus that will commit malicious acts. A virus that deceives us. A virus that leads the nation to surrender and say We give up. We give up. We give up. Though it was unfathomable, it was also vulnerable. For nothing is indestructible. In the end, it was defeatable. Such a virus is considered the virus that will lead people to compete for toilet paper. And I don't know why. The virus led people to work from home, and they discovered that it is better than working in the office. It's the virus that is so deceiving, and nobody knows who the culprits are. Nobody knows who the hosts are. Nobody knows who will live. Nobody knows who will die. A deceiving virus indeed. A deceiving virus indeed.

In return, it killed millions, changed lives, united us, and divided us. The virus is also politicized, very diplomatic, and deceptive. It chooses everyone.

It chose us humans.

So, what happened in that Wet Market? What began as it? What made the virus conquer the world? Yes, it conquered the world. The most advanced human civilization kneeled on it.

The virus was just another virus. But, of course, a virus always has a host. Of course, rare animals always come first. Or it could be common animals, like your dog, too. It just needs to do something. Deceive.... Imitate.... Replicate.... Mutate. That's the key to a virus that doesn't have to be deadly but can deceive the most protected and healthiest host. It can destroy them. They may survive. But some will not. Some will die in a week. But some will die in a day. So, very deceiving indeed. Is it a virus duplicating, imitating, or mutating in our gene DNA or RNA? It's so deceiving indeed.

We humans haven't seen such a virus for a very long time. The only good thing about the virus is that it did not kill a billion people. That's the only good thing about the virus. That means it is not deadly at all. It isn't as lethal as we thought it was. That's because, as assessed, there are more asymptomatic than symptomatic. That means many live a good life while having the virus, but only a few become sick. And good lord, only very few die.

So, let's assess the Wet Market in Wuhan. It's a bustling market. Full of products like goods, fruits, vegetables, farm animals, common animals, and, of course, the ones that are the real culprit, the rare animals. The ones that carry the virus that is probably not dangerous. The only problem is that there are just too many of those rare animals. It could be a pangolin. It could be bats. It could be any animal at all. So, what happened? A virus is not usually airborne. But, of course, there are airborne viruses. The ones that we cannot fathom. The ones that are not going to make us happy. That is because it can quickly get into us. Of course, the usual target is the nose. But the rare animals have been in that wet market for a long time. It did not do anything wrong. People were alive and happy. So, what happened? From what I assessed, the airborne viruses decided to find a human host, with multiple viruses present. Multiple viruses decided to replicate, mutate, and, of course, try to find another host. It will not do that in a day. It will not do that in a week. But it did. The virus was identified as COVID-19. The perfect virus that deceived the human being – all of us.

So, how did it conquer the world? A common airborne virus will try to reach the host and cause damage immediately. It can do that slowly. It can do it fast. The good thing is we can determine it. That's because we are all symptomatic of it. Of course, one of them is the flu. It always occurs every year. It continuously replicates and mutates every year. It replicates easily. The good thing is, we are all symptomatic of it. That means we can determine the origin and destination of the hosts. We know where it is going and can find a way to stop it. The best thing is we have the vaccine. That's it. We have the vaccine.

But COVID-19 is just too complicated. I never used the word "complicated" in my previous chapters. I always say complex. Damn it! That's concerning the virus. That's because we cannot see how it replicates and mutates. We don't know why there are asymptomatic cases and symptomatic cases. The worst is that we don't understand why some die.

The worst is that the virus is still around. Thank God we have the vaccines. But it is still around. Why? Even though the pandemic did officially end, the virus will not leave, vanish, or die. Of course, there are still a few who have it. The problem is they are fully vaccinated, healthy, and supposed to be immune, but they still get it. This is the part that I am apprehensive about. The virus is still replicating and mutating; one day, even with our best vaccines, it will become the Perfect Virus. A virus that cannot be stopped. A virus that will be immune to anything. A virus that will not surrender. That is what I am worried about. The virus that is so good only God can stop it. The problem is we are only humans. And, of course, we will not be immune anymore. I hope that day will not come. I hope that day will not happen. I hope that day will not cost a billion lives. No worries, that day will not come.

I hope...

To the family of the victims.
I mourn the death of your loved ones.
They will promise unrepeatedness

Chapter 47
Women Must Be Empowered

(They need more confidence)

The women have been struggling. Yes, they have been working. They have been pursuing equality. Yet, it's hard to get it. The majority of a company's executives are still males, as are the leaders of many governments, other high ranks, billionaires, and many millionaires. But the women persevered. They knew that was their only choice: to fight, to fight, and to fight for equality.

So, what is my goal? Yes, it is obvious. I want women to be proven equal to men. Women will turn out to be good leaders. They do. They are the real leader of a family. The community will do as well as a country. They are devoted to the benefit of society and the community. They have been proven to be reliable in becoming mothers in society.

For some reason, women are not treated as equals, just like animals. For example, most herds or groups of mammals are females. They knew how to work as a group. They knew they worked to provide nurture to their family, especially the young ones. They knew they had to choose the right male for their herd. What is the difference between human females and mammal females? Easy. The mammal females work in groups, but the human females do not. Usually, they work alone. As you all see, females are sociable. They work together and know how to socialize as females. But they never work together to make their life better.

What if we demonstrate that females need to work together as one? They can form a group of twenty to forty females working as one. If mammals, especially elephants, have matriarchs, then human females should too. There should be a matriarch in that group. We should create one. Who should be qualified? Of course, the smart one. We also have to include the experienced one. I guess an older woman who is already a grandmother should be the one. She should dominate. She should be perceived as the leader. Then, there will be other structures in the group. Like a nurturer for the young ones. A female who works to help the young kids.

As you all know, what women need is confidence. She can get involved with men such as a husband, father, or boyfriend. However, they cannot be reliable. She needs to get confidence from other women. A group of women should be trustworthy. They will provide a single woman with confidence. With that, will women be empowered?

So why empowered women? Why not? They are the central empowerment of the consumer economy. They spend more than men on anything except the kitchen sink. I apologize. Men do. But you know what I mean? Women tend to pay more. In Japan, teenage women or girls tend to be the biggest spenders. In the malls throughout the world, who do you see most? Women and girls. They tend to spend more. We need to be empowered. Provide them with more income and they will make a country's economy or macro-economy bewildered with consumption or consumer economy.

Women deserve to have more income. They will spend it, of course, in a good way. Not in gambling, drugs, alcohol, or everything else. But with everything else. We need to empower them. They have the right to be empowered in any way.

Again, they have the right to be equal to men. They deserve it.

Chapter 48
Billionaire's Paradigm (Greed)

(For it is not what it is worth, for it is all about achievement)

Indulgence. It is an act of satisfaction. It is an act of fulfillment. Greed. It is not about satisfaction. It is a desire to have more. It is addictive. It is undesirable. It is always in our core. It is almost a doctrine. It is insatiable. Appetizing. And most of all, an unstoppable trait.

Throughout the history of humankind, greed has been a habit. It has been a part of the culture of many civilizations. It is a core trait of the economy of undesirable poetry, for it created countries. It created empires. It conquered the world with no remorse. It conquered the world with no admiration.

Forgive me, for I am unjustified. Forgive me, for I have not dwelt with this diabolical trait. For greed is a Devil's companion. It is something that we cannot fathom anymore. We are now a society that can fight this together. We are a society of one.

I know you, the wealthy, cannot tolerate this. I know you cannot comprehend such a meaning. But you have to understand that it is still possible to cure this disease. Though greed is powerful, it can still be abstained from. Society can do it for you. You have to open your heart to them. And they will provide you with a satisfaction that is eternal, modest, and most of all, the power of humility.

For the Divine gave us the power to unite and to share. For the Divine gave us this gift of indulgence. And we cannot adhere to greediness.

So how do we overcome this trait?

Can we overpower greed?

Yes, we can. We, society, have always been in this undesirable consequence. We are never taught how to overcome this trait. That is because it is always on our minds. And our souls are burning with this undesirable malignant. I will teach you how to overpower greed. I will try to erase this from society. It will vanish, but we must do it thoroughly, for it is a matter of life and death. It will deal with our lives. It will deal with society. My teaching to overcome greed will not be violent. It will not be socialistic. It will be both liberal and conservative. It will be the only choice. Forgive me, for I cannot tolerate this anymore. Forgive me, for it must be changed.

This is just the introduction, and I will provide you with the real solution to overcome this greed and, hopefully, eliminate it one day. But as it is repeated, it is unvanishable.

Months later, I finally found a way to overcome greed. To overcome greed, we need faith, philosophy, the environment, possibly heredity or genetics, and most of all, discipline. To do that, we must pass on the technology, ideas, and knowledge I introduced in this book and see how beautiful the world will be with them. Will greed be overcome? Yes. The reason for that is that the Upper Class and the Middle Class will start to open their wallet for investments.

They are going to make money. A lot of money. The abundance of money will make it easy to obtain. It will be easy to overcome the financial loss. The loss of money will be minimized. Most people will win by earning more money. Greed will be overcome if we have an abundance of money. And that is the key—the abundance of money. Investments in stocks, bonds, real estate, cryptocurrencies, and many more will probably be abundant.

This will be my Last Supper.
You will be the one who will drink my blood and eat my body
You will find repentance and enlightenment
For I will die for your sin
(Christianity)

Chapter 49
(Set Pieces)

The Global Renaissance. World's Golden Age. The Global Rejuvenation. It is all part of a dream that may finally be realized in decades to come. My prediction is that it will occur in 60 to 80 years. It will occur exponentially. It will occur tremendously as we can finally achieve it. We can finally eradicate Global Poverty and have it vanish forever. However, preparation must occur. Set pieces must be in the correct position. To start the preparation process, we must begin with the foundation. The foundation is when we find a way to start an economy. For example, we must find a way to train or teach individuals or groups to start a foundation for society. We must provide schools, training facilities, academies, colleges, universities, and other institutions to start a community. Then, once the training or teaching is completed, we must find a way to provide those students with the given job opportunities. Of course, entrepreneurs must start a foundation to start businesses. The government must begin the process of governing. Laws, rules, and regulations must be provided, among other things.

These are the typical economic foundations. The goal of this foundation is to provide jobs and to start a middle class. A typical economic foundation. For some countries, it is an economic revival.

Yet, these are not the real issues of this chapter. It is just a basic process to start an economy, and that is the key. We must find a way to start an economy, macro or micro. We cannot just start an economy; we must find a way to ensure its continuation. But this chapter is not about starting an economy. It is about preparing to implement ways to tackle Global Poverty.

Belt and Road Initiative. The Chinese government provided an initiative to provide investment to countries in desperate need of roads and infrastructure. This is a program so successful that 145 countries are currently members. So, what did it do? I do not usually like Chinese principles, but this one made me jealous that the United States and its allies can implement them easily. The United States easily created one that is similar – Build Back Better World (B3W). The initiative seeks to address the $40 trillion worth of infrastructure needed by developing countries by 2035.

So, what is the reason I included these initiatives in this chapter? The Belt and Road Initiative is so successful that countries with new roads and infrastructure are finally beginning to experience an economic revival. However, the United States must implement one now. We are losing influence in developing countries due to the infrastructure projects made by the BRI. We must fight back. Build Back Better World or B3W will be a game-changer. It will work well if we provide it as a competition against the BRI. That is the only way it works. This is the only way for clarity. Developing countries have been asking for funds for many decades. China got to them first. It does not mean it is already too late. We must compete as well, for this initiative provides huge economic enlightenment for developing countries. It will also give a sense of accomplishment for the borrowing country, as these infrastructure projects will remain a testament to our achievements. We must do it now. Developing countries are receiving numerous economic incentives due to their confidence that one day they will reach the Renaissance Age. Africa, South Asia, South America, and the Caribbean are accumulating significant capital due to economic growth, except for countries affected by inflation, the Ukraine war, and other factors.

Infrastructure Project Initiatives are not the only factor in achieving a Renaissance; providing financial support, such as Basic Income, is another factor that may lead to it. But this is extraordinarily complex. Developed countries might thrive with these principles, but developing countries may face headwinds due to government uncertainties, such as Graft and Corruption. But Basic Income must be implemented. It is the only way for all people to accumulate such income. Basic Income does not need to be big at all. In the United States, you must provide a Basic Income of less than $1,000 per month to achieve fruition. You may need less for a Developing country such as the Philippines, which may only need approximately 10,000 pesos. However, it must be implemented well to ensure that the Renaissance will be accomplished. That will be in the next chapter called My Implementation (Execution).

Another factor that must be included in this Set Pieces chapter is the implementation of International Development Banks such as the World Bank, International Monetary Fund (IMF), New Development Bank by the BRICS, Asian Development Bank (ADB), and the Chinese version, which is the Asian Infrastructure Investment Bank (AIIB). There are many others, but these are the major International Development Banks that provide help to many needy countries. The World Bank and the IMF were established when World War II was almost over, based on the Bretton Woods Agreement, which was the idea of John Maynard Keynes. I am a Keynesian myself. I would love to provide more ideas from John Maynard Keynes, who single-handedly addressed the many issues of the Great Depression and the rebuilding efforts after World War II. He nearly created a perfect system of Keynesian Economics from Post-World War II to the 1970s, when the Great Stagflation occurred due to the actions of the Organization of Petroleum Exporting Countries (OPEC). But this is not about him. It is all about the Great Renaissance. If John Maynard Keynes were still alive today, he might have already provided an alternative to what I call the Great Renaissance due to his cohesive and abstract ideas that cultivate us all. He may already be surprised that today's society is on the verge of it. As I mentioned before, the Great Renaissance may be experienced in 60 to 80 years. It will be the next great Civilization for that civilization will experience an unimaginable outcome. But that will be in the next chapter of "Vision."

Another factor that must come to mind is the preparation of food, farm products, and many others. Once the Global Renaissance is achieved, food consumption will be in high demand, and new farmlands will be necessary to meet the supply needs of the upcoming middle class. New farmlands cannot be made into regular plains or cut-down forests. It must be implemented in a place that is uninhabitable but will still work with genetic farm goods. Two come to mind, and these two will work perfectly for the world's middle class, which will be needed at its fruition.

Sahara Desert and Siberia.

The Sahara Desert is a vast area that needs to be utilized more effectively. It must be cultivated and fertilized. New farm products today can be implemented in a desert environment due to their genetic modification. The Sahara Desert is vast. It lacks water. It lacks good soil. That is good enough. New farm products will be used on an extensive scale. We must ensure it will work, as many factors could lead to its demise, such as human abuse and natural disasters (drought, desert wind, etc.). Just 10% of the Sahara Desert is sufficient to supply the entire global middle class, and that is considered adequate. To achieve it properly, artificial rivers and lakes must be implemented. Weather medication, such as artificial rain, is needed badly. It can be achieved. It must be implemented.

Siberia. Another vast plain. It is just too cold, and there is too much snow. How to implement farms in such a harsh environment? Easy. We need to implement High Technology Greenhouses. Greenhouses that can withstand harsh conditions such as feet of snow, cold winds, and frigid temperatures are required. Like the Sahara Desert, Siberia is a vast area of land, an enormous expanse. A vast number of greenhouses can be implemented. This is a necessary job that must be done. This is the right way to do it.

Another contribution to the Global Renaissance is Nuclear Fusion. It must be implemented. Currently, the most outstanding scientists in the world are struggling to contribute more to technology than to input. The United States was the first to implement such an outcome. However, due to the cost of such technology, it may not be feasible for a long while. The technology was once thought to be easy to implement. However, as it becomes more complex to innovate, they discovered that it is more challenging than ever before. So why continue it? Why does it need to be implemented? Why must it be done? There is one reason. It is an output that empowers the Sun and countless stars in our universe. The output, if perfected, will be an infinite amount of energy. It will be an endless amount of power. And more effectively, it will be clean, renewable energy. It will take another five to ten years before such technology will be implemented. It is just too complicated with the Americans, Europeans, Chinese, and Japanese competing to see who can implement such technology. Americans were able to provide more output than input on such technology. However, they only did it in a few minutes. It needs to be an infinite amount of time, of course. That is the course that disrupts the scientists. It is just too complicated.

Another technology that needs to be prepared is the more

Advanced Desalination Plant. With almost infinite amounts of water, the sea and oceans are the ideal sources for water. However, due to the accompanying salt, which is challenging to manage, a Desalination Plant is the key to its removal. Desalination plants have been around for an exceptionally long time. However, due to the costly process of removal, a modification or innovation is needed to provide a feasible way of removing salt. Such an approach has been tried several times with no luck. However, once we find the right way of desalination, it will change our way of providing water to the people. The right way to provide water to the world. It is the only key because rivers, lakes, and other fresh bodies of water are not abundant or suitable enough to provide the correct approach.

These processes, approaches, techniques, and technologies are some of the preparations to approach the Global Renaissance. Such preparations are needed to achieve a better life, a better approach to needs, and safer access to power.

During my reign, I have seen too many anomalies.
I hope that in your reign it will not only subside but also be diminished.
For Anomalies must be tackled with perfection

My life is my boundary.
I hope once you come into my life, our boundaries will be as one.

I shared with you my world.
As you share yours, I complete
Love and Perfection

Chapter 50
(Execution)

This is it. Time is of the essence. Time is in a trial. For executing is the key to reaching Global Renaissance. Towards the end, we will prevail. We will achieve what humans were not able to accomplish in the history of humankind. We must provide a careful pursuit. For the key to implementation, the chapter needs to provide the right way to execute perfection. Execute the right approach. Execute the right actions.

So, what are these executions that need to be approached? Most of these are programs and approaches provided by the government and not-for-profit organizations. These are the executions that need to be implemented for each individual and group. It must be perfected to approach the Global Renaissance. It must be perfected to provide the correct goods and services to society and the world.

The first execution is the vaccines. Implementation of specified vaccines is key to good health and immunity. Dozens of vaccines need to be injected into our bodies. However, some are desperately needed, especially for developing countries. One of them is the malaria vaccine. We all know thousands of people, especially children, die of Malaria. Vaccines are the only way. Thanks to our innovation, we finally have a vaccine that can fight this dangerous disease. Such an approach is desperately needed in developing countries like Africa, where a vast number of children die of Malaria every year.

Of course, we will always think of COVID-19. The most influential of all pandemics. The disease that brought the world down and affected its economy. The pandemic is dividing my nation, which is the United States. The pandemic is so sophisticated that you cannot see which people have it. You do not know which person is symptomatic or asymptomatic. It took a year to provide a vaccine. The worst is that the disease has mutated, so new vaccines are needed again to tackle it. So far, five vaccines have been administered, but some people are not immune to it. The disease still gets them because it eludes itself. I want to thank God that we finally almost achieved immunity.

Another disease that needs to be approached is Dengue Fever. Yes, the disease is like Malaria, and it kills thousands of children, especially in tropical countries. The disease only occurs during the rainy season when water is available for mosquitoes to lay their eggs. The more eggs there are, the higher the survivability of the mosquitoes. It is a highly complicated disease. If you get it, you will have a tough time tackling it. If you are a child, you have a high tendency to die.

These are the diseases that I believe have vaccines. Yet implementing it requires the right approach, good planning, and, most importantly, a workforce to tackle it.

The pandemic comes and goes. Yet tackling it remains complex and elusive. COVID-19, Hantavirus, Ebola virus, and many others are just some of the viruses in the world that are so dangerous and elusive that, even after several years or decades, they remain challenging to approach. We must remain vigilant because it will disrupt the great resolve. It will disapprove of the Global Renaissance. It is a catastrophe waiting to happen.

Forests – the lungs of the Earth, the breather of life, and its home. We have been deforesting the Earth since the Industrial Revolution. We have been ruining it. But for the first time in decades, we are reforesting it. In this decade, we finally achieved a positive sign in reforesting, as we have finally reforested the planet rather than deforesting it. With this approach, we can achieve the Global Renaissance. Reforesting requires extensive labor, and such labor necessitates significant movement of the workers. That is because you do not reforest a backyard; you must reforest mountains, hills, deserts, and plains. You need an army of labor to climb mountains and hills. You need labor to carry those baby trees. It is a hard approach, but in the end, you will finally find accomplishment. Once planted, those baby trees can take care of themselves. You do not need to water them. They will survive on their own. However, only one out of five trees survives in the wild. But that is good enough. It is a better approach than doing nothing at all.

Implementation of Renewable Energy: Wind and Solar. The cleanest way to achieve power is also the safest way. So, will solar panels and wind turbines really work? However, it is definitely still not the most feasible way of dealing with power consumption. That is because it is not that dependable. Not that feasible, well. It can be stored in batteries, but a few homes are only powered by it. But it is the cleanest approach. It can fight climate change.

The implementation of a universal basic income is another aspect that needs recognition. It is an innovative approach to understanding how humankind finally achieves substantial income, yet it still requires a significant amount through occupation. That aspect has been introduced to many Western countries with tremendous success. People can finally spend money, but they still need jobs to earn more, as the money they receive from UBI is just for spending. People must work to produce more savings.

So, the next approach will be to tackle Climate Change. The

hardest of all implementations is due to the ongoing bureaucracy and politics. It is a straightforward approach, yet due to government intervention, it is simply hard to tackle. But we must remain resilient because one day we may encounter climate change first before the Great Renaissance. We cannot do that. However, the other side has also been great in its approach to Climate Change. They provide good evidence that is not climate change by showing that the Earth itself is getting hotter, without any evidence that humans are behind it. Only nature. They provide proof that even at the time of the dinosaurs, the temperature was so hot that the Arctic and Antarctica had no snow or ice. Yet, that is their only answer to everything. They cannot provide anything else.

But we do. We have evidence of Climate Change. We have been tackling it for a long time, and we have found ways to resolve it. Yet anti-climate change advocates will always pick up a fight. So, what is the best way to fight climate change? One answer only. Let the right-wing and left-wing work together in Washington so they can finally find peace and find a way to tackle Climate Change. We cannot fight Climate Change with a divided government. But you do not have to worry about it yet because the factors contributing to climate change have not been implemented yet. Factors that contribute to climate change have not been destroyed. Plus, people are resilient. They can withstand floods, wildfires, droughts, and other natural disasters. Some are even used to it. People in the Philippines experience two dozen storms or typhoons every year. Their wooden houses get destroyed, yet their lives remain. They have a tough time surviving, but they will survive.

You cannot be small, for your knowledge proves boundless
Your accomplishment will prove fruitful for you, for the Universe

God gave us life
You must give him eternity.

Your body is not your true self
For your soul

Chapter 51
Geopolitics On The Horizon

(Politics of Supply and Demand)

In the world of Global Renaissance, we must know that we are going to have different perceptions of what will happen in countries around the world, especially the major countries. The ones that we must know is that geopolitics will not be the same. It will be colossally upgraded. The global macroeconomies alone will not only be enormous but it will be overwhelmed with so many fortunes that it will cause enormous wealth.

So, what's the reason for it?

We must go to the countries that were at that time with enormous wealth. Those countries' wealth is so immense that their economic influence is unstoppable, leading to a Global Renaissance. The accumulation of wealth will be so easy that poverty will be eradicated.

How does it work?

Simple. The billions of consumers who will drive the world

financial market and Main Street will be so unstoppable that governments and major businesses worldwide will struggle to cope with demand. There is so much demand that it will cause an enormous deficit for the supply side. Government and businesses must keep up, or it will disrupt the macroeconomic environment.

Will it work?

Of course. As I mentioned in the Set Pieces Chapter and Execution Chapter, enormous projects must be built, and significant conservation efforts must be fulfilled to keep up with the enormous consumer demand. So much demand that the lists that I mentioned in the Set Pieces chapter and Execution chapter are not even enough to ensure a balance of supply and demand. Other ideas must be fulfilled to ensure a continuum of Global Renaissance.

So, who will be the major economies?

First, the biggest economy at that time will probably be China. There is no doubt at all that China will not only be the biggest economy in the world but also the most powerful in terms of diplomacy, military, and influence. So much power that the rest of the world will not only be intimidated but will also fear it in such a deep respect. There is such deep respect that its influence will lead to peril in geopolitical terms. China's people will amass so much enormous wealth that the demand will be unfathomable and unprecedented. Because of that, China will be at the forefront of wealth distribution and accumulation. It will be the biggest factor in providing the Global Renaissance.

The following country will be India. The economies between the two will be close, and India may surpass China by the 22nd century. The country will possess significant wealth as well, but it will not be as substantial as China's GDP per capita. That is because India still needs to catch up, while China accumulated wealth well before India. However, it is still the second-biggest economy in the world. It will be so enormous that just like China, it will experience disruption in the macroeconomic environment. Like China, it will lead to disruption in the demand and supply of goods and services. However, its manufacturing prowess will still not be the biggest or the most influential because historically, India does not have strong manufacturing capabilities. However, the Services Sector will be the biggest in the world. It will be bigger than China. Like China, India will be unstoppable. So much power will be accumulated. So much influence will be asserted. So much wealth will be distributed. Like China, it will have more than a billion middle class, and they will ask for many things and better things. These two countries will generate wealth so unimaginable that it will disrupt and accumulate so much that wealth distribution will be easy.

The next on the list will be the United States of course. However, wealth accumulation will not be as daunting as it used to be. There will still be a significant accumulation and distribution of goods and services, but it will not be the same as before. That is because wealth accumulation will be slow or will come to a standstill. That is because, in terms of demographics, the United States will experience enormous decay. It will not be the same anymore, but people's happiness will still be fulfilled. Education prowess, especially in College Education, will remain significant, but there will be competition not only from China and India but also from the rest of Asia. The colleges and universities of Asia will be highly demanding. The United States will still be at the forefront in terms of innovation and technology. Creativity will still be there, but hiring talented people from abroad will be futile. That is because most of the talents will stay in their home countries. They will not be intimidated by money and wealth, as they will stay in their home country, which will provide them with wealth that can compete with that of the United States.

The European Union will be next on the list. The Union has been at the forefront of the economy since the Renaissance Age. The countries comprised have fundamentals to ensure economic growth for developing countries around the world. France, Germany, and Italy are members of the G-7 Nations, whose power and influence have shaped Western Contributions that have cultivated the global economy since the industrial age. Besides the United States, the EU was the primary contributor of economic power to the entire world. However, recently, their economies have not contributed well, which will lead to the Union's dormancy for several years. The idea of the Union was once seen as a unified power. However, with the Euro currency not doing well and the stagnant economy not contributing either, the dormancy continues. If the E.U. needs to survive, it must cultivate, enhance, unify, and find ways to fix its demographics. It needs to build an industrial base in the technology sector to ensure its survival for years. One idea that I must present to the EU is to make a city where people work together. The city must provide a unified European culture to enhance the lives of different people from different countries. The main goal for that city is to provide a network of collaboration to improve ideas, creativity, and innovation, aiming to compete with Silicon Valley in the United States and Shenzhen in China. For that, the city will not only flourish. It will be a model of a new European economic power. The growth provides an actual united Euro.

The next on the list will be the African Union, or AU. I must put it on the list because it will be the next most dynamic continent in the world, and its power and influence will be felt in the 22nd century. Demographically, AU will be the youngest of all group countries or continents. It will experience the fastest GDP growth in the world, but it will be feasible as well. Wealth accumulation will be quick not only because of consumer demand but also due to the enormous natural resources. There are so many resources that the continent may produce the most crude oil and natural gas in the world.

Diamonds, gold, silver, copper, and many others will be on the number one list as well.

The last one on the list will be the rest of Asia. That includes the South Asia region, the Middle East, Southeast Asia, and East Asia. The accumulation and distribution of wealth from these countries will lead to another imbalance in supply and demand. Asian countries are among the most dynamic in the world, or the most dynamic that have ever been made or accomplished. Such countries provide labor that is hardworking, relentless, and resilient. Though Asian countries provide goods and services in fast-paced, resilient environments, their systems still prove amateur, especially in areas such as South Asia and ASEAN. But others are more advanced in providing goods and services that are fast, high-quality, and efficient. These countries are in East Asia and the Middle East.

So why are they last on the list?

Not only is it diverse in culture and politics, but the poor demographics are the ones that will disrupt economic growth. Not all of them will experience it, but mostly the East Asian countries like South Korea and Japan. Singapore and Taiwan(?) may experience it as well. Old age will take a toll on the economic growth of these countries. Japan will experience it first. Others will follow. Unless the government intervenes, it will cause a significant economic disruption that may even cause an economic slowdown or even a mild recession. Singapore is trying, yet even with special programs and tasks in hand, it still has issues propagating its population. Taiwan, South Korea, and, of course, Japan are having the same problems as well.

A flightless small bird has more freedom than a toothless fox

Chapter 52
(Conclusion)

A luminary. That is what I am. A person who brightens people. A person who signifies others. A person who lives to the fullest. A luminary does not appreciate money. I want to experience good outcomes. I want to experience happiness and inner peace. It is not about who I am but about what I did. But this is not about me. It is all about my vision—the last chapter of this book. The chapter deals with the outcome—the final piece.

Global Renaissance deals with grandeur. It deals with the World's Golden Age. It deals with the outcome of happiness for all humankind. So, what is my ideal vision? My grandeur. My happiness.

My first vision is a world without poverty. With poverty gone, the outcome is finally feeling unrustable. For me, it shines once I see it. I finally felt unboundedness. Poverty alleviation is one of the most insignificant programs that humans have experienced. It is just too hard to overcome. It is just too hard to comprehend. It is too hard to appreciate. I felt a hole in my body, for that hole signifies unreality. That means I never felt perfect in my body because that hole feels like my gap about poverty. Tremendous may it seem, I feel disgusted, yet I must enlighten myself, and I finally did. Sixty to eighty years from now, I might be dead, but I will see it from heaven itself: the Global Renaissance and Its True Meaning.

God gave me grace. God gave me the outcome. God gave me significance. I finally found happiness because the implemented technology has revealed Global Renaissance's true nature. With innovative technologies, jobs have been found and implemented. Businesses have been made, and profits are set aside. Actual progress has finally been achieved. Governments around the world have finally achieved a budget surplus as their income has appreciated significantly. Individuals and families have finally achieved better lives because their income has increased, but their jobs will also be longer. Peace and prosperity for all have finally been made.

Space - the last frontier and the final one. The previous prominent contribution of humankind will finally be achieved. Exploration has finally been completed. Now, space's natural resources can finally be mined. And there are many. More like infinity. For these reasons, we will have a considerable number of resources on this planet, and mining will not harm our natural environment. We can finally go to Mars. We can finally go to the moons of Jupiter, Saturn, Uranus, and Neptune. We can finally check the Asteroid Belt, Kuiper Belt, and the Oórt Cloud. And by the next century, it will be the Proxima Centauri – the closest planetary system to our solar system.

Power can finally be achieved universally, infinitely, and cleanly. With nuclear fusion, it can provide such a significant amount of power that we can finally plug in as many gadgets and appliances as we want. An infinite number of homes will be powered by it. With limitless power comes infinite responsibility. We must be responsible for what Nuclear Fusion will bring to our society and the outcome it will produce. For now, no destruction has yet been achieved by Nuclear Fusion. Yet we really do not know if it can accomplish that. Nuclear Fusion is still in the experimental stage, and it is difficult to generate electricity from its core. All we know is it is clean, but we do not know that until it is finally in an operational phase. However, promises are already being made and nuclear fusion one day will replace other forms of power sources such as Nuclear Fission, Geothermal Power, and many others as well.

Shortage of water is now a myth. Water abundance is now more common. For Desalination is the key to the freshwater supply. The thought of having water transformed from salt water to fresh water is significant but it endorses significant expense due to the techniques being used. To gain a simple transformation of salt water to fresh water, a modification is needed. As I mentioned before, desalination is the key to producing fresh water due to the abundance of salt water. If we can simply perfect the technology, then it is possible to eliminate water shortages due to drought, war, and many other causes.

So, what is the reason for all these causes? What are the implications? If possible, what will be the result? My vision is not just to signify the Global Renaissance Age but to prolong it. For this is the only key for everyone to find true happiness and harmony.

God has given us the freedom not to compete but to cooperate. We live not to conquer but to give peace and prosperity. We were once animals. We are humans now. We live to pursue heaven and in heaven we found enlightenment. This is our age. A significant age to pursue for we are near the global renaissance. It will not be in our generation, but it will be the generation that is perfect.

I was once in a dormant phase too, but I must pursue closure. For what I want and what I need is to help eradicate the poverty of this world. I just cannot comprehend why the world has a ginormous wealth inequality. I just do not know why people in poverty exist and there is nothing we can do about it. For that, my main forefront is to tackle poverty alleviation. I must because it must be done. I must because there is nothing else. I must because it is my only choice. It is the fundamental key to true freedom, happiness, and enlightenment. For that, I must fulfill this goal. My only goal. The goal is to realize a true Utopia.

MY DREAM WILL NOT BE SIGNIFICANT WITH THIS BOOK, BUT OTHERS WILL PURSUE IT. IT IS JUST THE BEGINNING, BUT OTHERS WILL PERFECT IT. FOR THEY WILL PROVIDE A BETTER PIECE. THEY WILL PROVIDE THE NEXT CHAPTER. BUT I HAVE TO GUIDE THEM. AND THEY WILL GUIDE THE OTHERS AS WELL.

SEE YOU IN THE NEXT CHAPTER....

For we wonder in fear... we must pursue infinite encouragement... for we are humans... we are not perfect... but we are boundless... the true beauty... the true wonders... of the world.

God himself can only guide you.... You must pursue the true meaning of dreams and happiness... for that will provide true enlightenment... a true pursuit to see and meet God himself.

And in the end, we are all just mortals. But great ideas and creativity create an eternal, immortal, and infinite entity.

Citatio Finalis

As BRICS influences the world, the West
needs to exert influence as well, for it is dwindling.
It is diminishing.
BRICS now has a presence in most of Asia, Africa,
and Latin America. The West must foster friendships
with those continents and with BRICS as well.
For it is the only choice they have, or it is too late.

Part Two
My Miscellaneous Poems

Introduction

I write poems as a hobby, and I have written more than a dozen of them. The majority of them are in English, but there is one that is in Filipino or Tagalog. I decided to put them in my book just for your entertainment. Just an FYI, some of them are revolutionary.

The first poem I will provide you is written in Filipino or Tagalog. This is the revolutionary poem I am talking about. I will give the Filipino version first, and then it will be translated into English. Check it out.

Chapter 53
Dahil sa Ako ay Hampaslupa

Ako man ay isang walang katupad
Walang trabaho at walang pera
Kung ituturing man akong wala
Hindi ako mabubukod
Hindi niyo ako magagalig

Dahil ako ay isang hampaslupa
Hindi niyo ako mawawasakin
Dahil ako ay may galang
Ako ay may disiplina
Ako ay isang tao din

Dahil ako ay isang hampaslupa
Ako ay lalaban din
Dahil hindi ako madaling matalo
Ako ay Pilipino
Pilipino ang aking angkin

Dahil ako rin ay pinanganak na inosente
Inosente ang paglaki ko
Inosente din ang pagkapatay ko
Dahil hindi ako mamatay tao
Susundin ko ang batas kailan man

Hampaslupa ang paglaki ko
Hampaslupa ang pagkamatay ko
Pero sa tingin ng Diyos ako ay hindi hampaslupa
Dahil sa totoo ituturing niya akong mayaman
Ituturing niya akong maginoo
Ituturing niya akong totoong tao

Dahil hindi ako hampaslupa
Mayaman ang loob ko
Mayaman ang anyo ako
Mayaman ang aking pananalig
Mayaman ang pagkatao ko

Bugbugin na bugbugin niyo ako
Tatayo ako
Pahirapan niyo ako ng pahirapan
Hindi ako lulubog

Wala man kaming pera marami kami
Kailangan lang naming magkaisa
Dahil lalabanan ka namin
Matalino din kami

Ang laban namin ay hindi pisikal
Lalaban kami gamit lang ang salita
Walang Patayan, Walang masusugatan

Walang masasaktan
Dahil matatalo lang kami sa ganyan
Matatalo lang kami

Sa huli kami ang mananalo
Sa huli kami ang matagumpay
Sa huli mararanas naming ang utopia
Sa huli mararanas namin ang pagbabagong-sibol

Oo, ako ay isang hampaslupa
Wala akong pake

Chapter 54
Because I am a Poor Person (English Translation)

I, too am a person
No job and no money
If I am considered nothing
I will not be excluded
You will not disturb me

Because I'm a poor person
You can't destroy me
Because I am respectful
I have discipline
I am a real person, too

Because I'm a poor person
I will fight too
Because I'm not easily defeated
I am Filipino
Filipino is my origin

Because I, too, was born innocent
I grew up innocent
My death will also be innocent
Because I am not a killer
I will always obey the law

I grew up on the ground.
My death was a disaster
But God thinks I'm not worthless
Because in truth, he will consider me rich
He will consider me a good man
He will treat me as a real person

Because I'm not a poor person
I am rich beneath me
I am rich in appearance
My faith is rich
I am rich in character

Beat me up
I will stand up
Torture and torture me
I will not sink

We don't even have money, but there are a lot of us
We just need to unite
Because we will fight you
We are also smart

Our fight will not be physical
We will fight only with words
No Killing, No Injuring

No one will be hurt
Because we can only lose like that
We will only lose

In the end, we will win
In the end, we are the successful ones
Ultimately, we will experience utopia
Eventually, we will experience rejuvenation

Yes, I am a poor person
I don't care

I also wrote other poems. Please check them out. I hope you will like them. Here they are.

Chapter 55
Barbarism

It mitigates me
For your splendor comes with an atrocity
The beauty within
It distresses for it is just an illusion
You have the power to destroy me
Yet the power you wield could not even touch
my hair
You surround me with your influence
For your soul has the utmost beauty
Yet the influence of nothingness on me
You seem diligent
But you are not that brave
You are not a warrior
Nor my companion
You are just a barbarian
A barbarian who doesn't respect anyone
Nor collaborate
Nor friendly

You are just nothing to me
For I am a woman with the utmost beauty
And my beauty doesn't linger on you
It doesn't linger on me
For my husband was killed by a sword
But my heart will never be yours
Nor my body and soul
You are barbaric!

Chapter 56
Afraid to Get Near You

For I am afraid to be with you
I know it's awkward
I know it is unimaginable
For I see you, I died in a cat's nine lives
My heart is pure
Yet my mind is nervous
My body cannot concentrate
My soul is frozen
My head is struck by lightning

Yet I must be confident
I must be with you
I must love you
I must penetrate your mind
I must unfreeze your mind and soul
For my body is now as hot as hell
For my heart has opened
I am now a seven-wonder

I am now in a thunderstruck
I am now benevolent
I am now pure
For approaching you
The confidence now bestowed
I think I can conquer you
Just give me immortality
Give me confidence that cannot be stopped
For now, I believe I can be with you
I must believe
I must adore
I must love....

Chapter 57
Beyond my Redemption

I was once in a palpable state
Forever and beyond, I cannot undermine
The true meaning of life
The true meaning of my surroundings
The true meaning of redemption
For I am beyond the capability
Full of malaise
Full of naiveness
Full of disaster

I eat to ensure survivability
Yet I cannot fathom it anymore
I am beyond irresponsibility
Beyond vulnerability
For I am not easy
But I am weak
I need redemption
I need a new life

I need a new surrounding

You came to my life to enhance it
You came to my heart to purify it
You came to my soul to rejuvenate it
You came, and I became redeemed
For you are my life
You are my love
You are my soul
And I become one because of it
I have become a different person
For I am the light
I am the redeemer
For now, I am just a nobody
Redeem me and I will be beyond you,
They and everyone else
For you are my love
You are my everything

Chapter 58
Broadening of the Mind

I cannot proceed while enhancing it
I cannot fathom what's going on with my mind
I cannot portray myself for I see is you
I need you, for you are the key to my world
The key to my mind
The key to my heart
The key to my happiness
Be with me, and I will give you everything
For I don't know what's going on in my mind when I see you
For I don't know why my heart beats so fast
And my mind is expanding like a magic will happen
But don't be afraid, for I am afraid to lose you
Losing you will turn my thoughts into dormant
Losing you will turn it into a disaster
Losing you will destroy me forever
For you are my only thoughts
For you are the reason I am living

You are the reason I am alive

So please help me
Propagate me
Turn me into a diamond
For I will conquer the universe
For I will conquer you
For I will conquer your heart
Your soul
Your love
Your eternity
Be one with me and we will be unstoppable
We will be immortals
We will be Gods

Help me... Help you...

Chapter 59
Calamitous Philippines

Order, as it seems, yet calamitous features must be restructured
Devastated as it becomes, it will still be the envy of the world
Togetherness is a possibility, but the division is a nuisance that can't just vanish
For the world needs it, the world needs its people
Smart as it seems, the discipline is just too weak, disastrous, and malignant
People together as one will be colossal, yet together as one in the structure of a monster will proclaim it as devastation
The Philippines is a magnificent country indeed, yet disaster is always a common entity
Because the disaster it caused is the attitude, personality, selfish individualism, and destructive qualities

Yet I see Filipinos as a future for the world, for they
are smart
Today's generation might still recover, but the next
generation must be focused, for they need
discipline
They need enlightenment
They need needs
Yet I see the future indeed
For Filipinos are the smartest, yet they need discipline
and knowledge
They need God
For the almighty will give them blessing, but to be
blessed is to be enlightened
To be one with Him
To be one with the world
They must have a Lee Kuan Yew moment

Chapter 60
For I am Just a Frog

For I leaped with no boundaries
Yet I do not land on the right path
For my path only led to my demise
For I am just a prey
A vulnerable prey
And to them, I am just a nobody
But I can smell their fear
I can sense their vulnerability
To my lead and to my wonder
I have seen everything
Yet my innocence only knows in front of me
For the world is hiding what is beneath
The world is hiding the truth
For those predators will eat me if I find out the truth

But I am no ordinary frog
I am a poisonous frog
Whoever eats me will be extinguished

Whoever eats me becomes my prey
For the truth, I am the real predator
And they will be my victims
For the world knows me
They know I am no frog

Chapter 61
It Has to Be a School

It is in the middle of the National Mall
Then it must be a Smithsonian's
It is near the Government Buildings
It must be a Federal Building
It is near the Embassy
It must be an Embassy as well

It is not even a school as well
Because once inside, you think it's a well-designed
office
It was probably designed by one of the best interior
designers in the world
It could be a museum as well because it also looks like
a museum
But why is it a school?

It is a school that redefines real estate
Location? Location? Location?

I don't think so. It wants to be seen as well
It wants to know that it is there
It wants to know that the students will engage
For the students are Graduate Students
Of International Studies
Students who want to know the world
Students who want to change the world
Students who want to carve the world

But it's not just about the students
It is about what it's all about
The education
It's Johns Hopkins anyway
They will fight for change
They will fight for progress
For people want to be engaged
People want to know
And of course, at the left window of that building
There is a building that always wants to be seen
The Capitol Building
Best view ever

One day, all schools will be in this location
Because it's the location after all
Probably not
Probably it's all about the habitat, not the environment
JHU-SAIS

Chapter 62
Megalomania

So vast... so vast the eye can see
So troubling for what I see I conquer
For my ego leads me to temptation
For my ego leads me to destruction
For my ego leads me to a calamity of never-ending

I feel invincibility for I feel a strength of a rock
I feel invincibility for I can fight monsters
I feel invincibility for I see inside your heart
For all I see is you

But life must go on
Life must not depart
For life is too short
Too narrow to be an eternity
For life has its ups and downs
Yet it is vulnerable to my ego
For my life is an ego

An ego that needs more
The more it needs the more desperate it is

I will live. I will conquer
For my mind leads to the light
But darkness keeps me from that light
Darkness keeps me from being a saint
For the devil spoke to me
And the devil provokes me to come to darkness
For that is my future
That is my fate

What will be in the end?
For my blood is now obliterated
What will become of me
For my life will be at an end soon
But it will be worth it
I will make it worth it
For the devil is there and he is worth of it
And he will lead nothing into temptation
But he will deliver
He will provoke
For God is still sleeping
He has not woken up
He probably does not even care

Chapter 63
Synchronous

Sync with me
Invigorate my mind
Collaborate with me
For my heart and soul are now the same
Mitigate me
For I am no longer bound to others
I am bound to you
For my faith is in you
So, sync with me

My devastation is delighted
For I am enlightened
I am your hope
I am your light
I am your victory
I am your demi-God
So, sync with me

Follow my rules
Follow my inception
Follow the true meaning of love
For the truth never reveals until it sees
Forgive me but I must see
For you are no longer ordinary
You are special
You are my light
You are my soul
You will enlighten me
You will become me

So, sync with me
And I will show the true meaning of love
The true meaning of life
The true meaning of worthiness
My life will be unparallel until I see you
Sees you will give me closure
Give me wonder
Give me an extraordinary experience
For I am now yours
I am now special

Sync with me
And I will prove to you my worth
For that will never break us apart
So, sync with me
And I will call it inseparable

Chapter 64

The Epitome of Shenzhen

The Epitome of Our Dreams

Is it remarkable?
Is it dynamic?
Shenzhen....
Once a small fishing village
It is now the Center of the World
And it will be the Center for years to come
For its people are the key
The key to progress
The key to its success
The key to its prosperity

Though people have no origins
All migrants from different places
All lost but remain confident about the future
The people endure hardship through hard work and
clemency
Through ideas and cooperation
Through trust and will of the government

Shenzhen is now the Epitome of the developing
world

From a small fishing village
Now becoming a super metropolitan like no other
Shenzhen is now a True Dream that has become a
Reality
For its legacy will endure new ideas and creativity
Its legacy will endure a high-class society
For the world is now in Shenzhen
For Shenzhen is in our Dreams
It will be our legacy for thousands of years to come

Chapter 65
The Manila's Descent

Does it prove to be on a backlog century
Does it prove to be fatal
Does it prove to be in a decent harmony
Manila.... The Pearl of the Orient
For the world is not worthy to be in its feat
For the world is not worthy to be in Manila's mind
Hypocrisy can be anonymous
But Manila is not anonymous
It is unique
It is a piece of a beautiful jewel that can only worn by Goddesses
A jewel that is priceless to the world cannot price it
It cannot determine the real meaning of the city
For its people are determined with precision
With hope and clarity
With joy and clandestine

I am now in a descending mode

Diabolical as it seems Manila can be proven to be an emotional
It was once a city of nothingness, but dynamism has opened it
Let me open myself up
Let me prove my worthiness
Let me prove my quality for it is high
For it is the truth
For it is real... it is real
Manila is now a strong entity
A diamond that resembles soul freedom
A purge that can't be unmade
Amid my insurrection
Amid the destruction the city faces
It did not only thrive, but it also became a world treasure
It becomes the mortal's Olympus
A Demi-God's paradise

As the economy immortalizes, Manila becomes a world-class
It bestows a true intimacy
Of a world-class society
Of the elites and the common people
Let me face it
Manila is a wonder that you cannot notice
It is just there
Really there
Therefore, I am now adhering
To Manila... a true Luminary
A true compass that guides the city into the new century

A guide toward a heavenly descent
A phrase that cannot be read
For the century is now capable of embracing the true
meaning of love
The true meaning of life
The soul of Manila burns brighter as it finally
descents
It has to for the city to be ready for the new century
The new century indeed for Manila you are finally
descending

Chapter 66
The Providence of Hong Kong

Will it ever be the same?
The countless ages of catastrophe that never come of precision
Such disaster disgruntles and unbalances the society that was once elite
Yet their eliteness has now been dissolved
Dissolved for such prowess of the opponent cannot be handled anymore
It all became diffuse
It all became precise
It all became a never-ending rain

For we live in a disaster
Such disaster cannot be comprehended
Cannot be fathomed
The emotional brokenness is still not dissolved
For once we are in Hong Kong
Now we are nothing

For once we are in Hong Kong
Now we are hurt
For once we are in Hong Kong
We are now unequilibrium

We live
We die
But Hong Kong will forever be Hong Kong
Forever be freed
Forever be indestructible

It was once the darling of Asia
None no more for we have been outcompete
For we were once the epitome of the world
None no more for we were betrayed
For we were once the hearts and minds of the
younger generation
It still is but slowly evaporating

We are Hong Kong
We are resilient
We are Hong Kong
We will fight
We are Hong Kong
We bet our lives on this city
We are Hong Kong

And forever we are
For we are one

Chapter 67
The Shibuya Fields

The field of vision, the field of glory
The field that encompasses wonders
The field that requires extra perception
This is Shibuya
This is Tokyo
I can relate Japan's world to this place
Yet I don't know if something must be done
Everything is perfect
Everything is a wonder

Tomorrow will always come
Yet I can see the future of this world as an
extraordinary
I can see the past
I am meeting the present
Yet the future will be engaging
For once there is Tokyo, Shibuya will always be part
of it

For once there Japan, God gave it a meticulous name
Yet Shibuya will always be in his memory
It will be in his body of nakedness
It will be his making

I am glorified yet petrified for Shibuya will be in my
presence
Yet I do not see the real meaning of it
Why am I here
Why I must see it
Something is going on?
Is it the society?
Is it a vision that is waiting to happen?
Or I am just a pest that cannot fathom the field
A true field that is vast with no boundary
And that is Shibuya
An unbound field
An urban field
Or just a field of nothingness
For there is no actual field
It is just a dream, a field of dreams
A field that only God can make

Chapter 68
Tomorrow Never Knows

For we can't predict what tomorrow knows
All we know is it will be hard to blow
The meaning of life and how it shows

I can't seem to see the real wonder the world provides
For what I can think of it seems to not abide
I don't see to blame the real meaning of life
For what is the meaning and what can thrive

Today, tomorrow, and yesterday will never
really show
The real prediction it can throve
It paralyzes me for it cannot be drove
The world of tomorrow the future holds

If I can just set the real meaning of life aside
The real situation will never subside

For I am a hero, but I never abide
The long sedition will bring nothing but tides
In the real world, the real flame will be nothing
    but fire

Chapter 69
Velocity of Light

Amendable...
    Immeasurable...
        Intolerable...

That's how the light projects itself
For it cannot fathom the matter
All it has is pure energy
And in pure energy is its immense power
So powerful it rejects time
So intolerable it rejects space
So immeasurable it rejects physics itself
For nothing can run better than the speed of light
For nothing can fathom such an undeniable course
For the course can only be seen in no time
For no one can see it
And in the end of the day
We will see how vast it is
How intolerable it is

How tremendous it is
But this is just nature
And the light will be natural no matter what
It will guide you
It will tolerate you
You will be astonished

Chapter 70
Why Sacrifice?

You made it regret the true principles of sacrifice
It is not done with negligence but with will and logic
It does not seem the result fits
For the man who dealt with you manipulated the system
So, you died in vain
And in vain justice was not done
The prelude of the method must not be done again
But I will provide justice
Once the time has come
I will provide the rights for it is the only peculiar way
To provide rights and to provide justice is to provide the true sacrifice
For the true sacrifice is me
And I will provide it to you
My life for your life
My love for your love

My soul for your soul
I will ensure justice
I will ensure the right one

Go and heal
I will wait for you

Chapter 71
Worthy

Prove me... Weigh on me... Check me
And I will prove you no wrong
Made me... Collaborate with me... Ask me...
And I will show you the truth of me
Find me... Navigate me... Invigorate me...
For I am everywhere but I am nowhere

Once no worthy... Once no shame...
My mind has opened but gloomy and murky
Once found... Once seen
I will show you the true power of the mind
Once lifted... Once corrosion is gone
You will find no dirt for it is clean
Everyone knows... Everyone will take responsibility
As my responsibility is for the world

Don't you cry... For I will always be alive in your
heart

Don't you fight... For that proves unworthy
Don't you be mad... For that will leave into emptiness
Don't you fall... For that will lead to nothingness

May the Lord guide you from your heart, mind, body, and soul for you will be enlightened
You will be worthy
I will place myself as a big part of the society
I will place myself as who is the one
And once accomplished... I will be bound to the world

Bibliography

[1] China Economic Quarterly Q1 2023 (2023). PricewaterhouseCoopers. https://www.pwccn.com/en/research-and-insights/china-economic-quarterly-q1-2023.html

[2] Revising 2023 GDP data not to impact economic growth in 2024: NBS official (2024 December 28). Xinhua. https://english.www.gov.cn/archive/statistics/202412/28/content_WS676f5abac6d0868f4e8ee562.html

[3] World Factbook by the Central Intelligence Agency (2024 May 21). Central Intelligence Agency. https://www.cia.gov/the-world-factbook/countries/united-states/#economy

[4] Textor, C. (2024 September 20). Degree of urbanization in China in selected years from 1980 to 2023. Statista. https://www.statista.com/statistics/270162/urbanization-in-china/

[5] Kharas H. and Ren J. (2024 January 22). 'The Rise of the Global Middle Class': An interview with Homi Kharas. Brookings. https://www.brookings.edu/articles/global-middle-class-interview/

[6] Chinese Civil War 1945–1949. (2024). Brittanica. https://www.britannica.com/event/Chinese-Civil-War

[7] Browne R. (2024 October 16). ASML just gave us a first glimpse into how U.S. chip export curbs will dent its China sales. CNBC. https://www.cnbc.com/2024/10/16/asml-2025-outlook-shows-us-chip-export-curbs-impacting-china-sales.html

[8] Creighton J. (2019 August 21). The Kardashev Scale - Type I, II, III, IV & V Civilization. Futurism. https://futurism.com/the-kardashev-scale-type-i-ii-iii-iv-v-civilization

[9] Siegel E. (2021 December 10). Ask Ethan: How Large Is The Entire, Unobservable Universe? Forbes. https://www.forbes.com/sites/startswithabang/2018/07/14/ask-ethan-how-large-is-the-entire-unobservable-universe/

10 Baird C. (2022 July 20). Is a black hole a 2D or a 3D object? West Texas A&M University. https://www.wtamu.edu/~cbaird/sq/2022/07/20/is-a-black-hole-a-2d-or-a-3d-object/

11 Lea R. (2024 December 7). How big can 'ultramassive' black holes get? Scientists may have the answer. Space.com. https://www.space.com/ultramassive-black-holes-how-big

12 Black Hole Basics. (2024). NASA. https://science.nasa.gov/universe/black-holes/

13 Planetary Nebulas (2024). Center for Astrophysics of Harvard and Smithsonian. https://www.cfa.harvard.edu/research/topic/planetary-nebulas

14 Lea R. (2023 May 4). If dark matter is 'invisible,' how do we know it exists? Space.com. https://www.space.com/if-dark-matter-invisible-how-do-we-know-it-exists

15 Mortlock D. and Turner E. (2001 October 21). Gravitational lensing in modified Newtonian dynamics. Royal Astronomical Society by Oxford Academic. https://academic.oup.com/mnras/article/327/2/557/1040696

16 "Has a new study shown DARK ENERGY doesn't exist?!" *YouTube* Uploaded by Dr. Becky (2025 January 3). https://www.youtube.com/watch?v=T6BKDhoGTHE

17 The strongest evidence for a Universe before the Big Bang. (2023 March 16). Big Think. https://bigthink.com/starts-with-a-bang/evidence-universe-before-big-bang/

18 King B. (2016 December 9). Why does Siberia get all the cool meteors? Phys.org. https://phys.org/news/2016-12-siberia-cool-meteors.html

19 Flatcreekindev. (2017 December 30). The Bison: from 30 million to 325 (1884) to 500,000 (today). Flat Creek Inn. https://www.flatcreekinn.com/bison-americas-mammal/

20 The Passenger Pigeon. (2001 March). Smithsonian Institutions. https://www.si.edu/spotlight/passenger-pigeon

21 Hecht J. (1993 February 13). Science: When dinosaurs roamed Antarctica . . . New

Scientist. https://www.newscientist.com/article/mg13718603-300-science-when-dinosaurs-roamed-antarctica/

22 Kates-Harbeck J. (2010 November 28). Magnetic Nuclear Fusion. Stanford University. http://large.stanford.edu/courses/2010/ph240/kates-harbeck2/

23 Cherry K. (2024 April 22). What Are the Gestalt Principles? Very Well Mind. https://www.verywellmind.com/gestalt-laws-of-perceptual-organization-2795835

24 Jacobs F. (2024 June 24). Hurricanes never cross the equator. Here's why. Big Think. https://bigthink.com/strange-maps/hurricanes-versus-equator/

25 Rare South Atlantic Tropical Cyclone. (2004). Earth Observatory. https://earthobservatory.nasa.gov/images/12935/rare-south-atlantic-tropical-cyclone

26 Davidson J. (2014 December 19). Cuba's Classic Cars May Be Available, But Are They Worth Anything? Money. https://money.com/cuba-classic-cars/

27 Toque E. (2022 April 4). House/Apartment Prices in Cuba: Where Is it Cheaper to Buy? Havana Times. https://havanatimes.org/features/house-apartment-prices-in-cuba-where-is-it-cheaper-to-buy/

28 Trindade G., et al. (2024 May 30). X-ray induced degradation during XPS analysis of dicarboxylic acid powders. Applied Surface Science Vol. 656. Science Direct. https://www.sciencedirect.com/science/article/abs/pii/S0169433224004161

29 Olson S. (2016 September 16). Shenzhen: From fishing village to global metropolis. Hinrich Foundation. https://www.hinrichfoundation.com/research/tradevistas/fdi/shenzhen-foreign-direct-investment/

30 Aung San Suu Kyi: Myanmar democracy icon who fell from grace (2021 December 6). BBC News. https://www.bbc.com/news/world-asia-pacific-11685977

31 Aung-Thwin M. and Aung M. (2025 January 4). Resources and power
In Myanmar in Economy. Brittanica. https://www.britannica.com/place/Myanmar/Resources-and-power

32 Miller G. (2023 December 13). Panama Canal transits plunge as larger ships are turned away. Freight Waves. https://www.freightwaves.com/news/panama-canal-transits-plunge-as-larger-ships-are-turned-away

33 Thai – Italian Chamber of Commerce. (2021 November). Land Bridge: The Next Regional Maritime Center, the Connecting of Andaman and Gulf of Thailand. Issuu. https://issuu.com/thaitch/docs/nov_2021/s/14215200

34 Turner J. (2019 October 24). Destination Singapore: behind the rise of the world's top shipping centre. Ship Technology. https://www.ship-technology.com/features/why-is-singapore-port-so-successful/

35 Chipchura M. (2023 January 30). 10 Busiest Shipping Ports in China. Advanta Sourcing. https://www.advantasourcing.com/blog/busiest-shipping-ports-in-china.

36 More white-collar jobs available in SG vs. other Asian markets: report (2024). Singapore Business Review. https://sbr.com.sg/hr-education/news/more-white-collar-jobs-available-in-sg-vs-other-asian-markets-report

37 Mahbubani K. (2015 August 4). Why Singapore Is the World's Most Successful Society. Huffington Post. https://www.huffpost.com/entry/singapore-world-successful-society_b_7934988

38 China's South-North Water Diversion Project: Navigating progress and challenges. (2024 February 26). Aqua Tech. https://www.aquatechtrade.com/news/water-treatment/china-south-north-water-diversion-project

39 Zimmer K. (2019 August 28). Why the Amazon doesn't really produce 20% of the world's oxygen? National Geographic. https://www.nationalgeographic.com/environment/article/why-amazon-doesnt-produce-20-percent-worlds-oxygen

40 Johnson M. (2024). 5 interesting facts about the Pantanal, the world's largest tropical wetland. World Wildlife Fund. https://www.worldwildlife.org/stories/5-interesting-facts-about-the-pantanal-the-world-s-largest-tropical-wetland

41 Stephenson A. (2024). Lake Mead Water Crisis. AypoTech. https://www.aypotech.com/career-central/lake-mead-water-crisis

42 Fearon R. (2022 September 13). Building Atlantropa: One Man's Plan To Drain the Mediterranean Sea. Discovery. https://www.discovery.com/science/building-atlantropa--one-man-s-plan-to-drain-the-mediterranean-s

43 Kirsch C. (2020 December 30). State of Jefferson: The Three-Time Failure.

Medium. https://chriskirsch.medium.com/state-of-jefferson-the-three-time-failure-b5618413c43e

44 Mukherjee S. (2023 February 23). Nalanda: The university that changed the world. BBC. https://www.bbc.com/travel/article/20230222-nalanda-the-university-that-changed-the-world

www.ingramcontent.com/pod-product-compliance
Lightning Source LLC
LaVergne TN
LVHW090309210926
848480LV00001B/179